Cold Tolerance in Plants

Shabir Hussain Wani • Venura Herath
Editors

Cold Tolerance in Plants

Physiological, Molecular and Genetic Perspectives

Springer

Editors
Shabir Hussain Wani
Mountain Research Centre for Field Crops
Khudwani, Sher-e-Kashmir University
of Agricultural Sciences and Technology
of Kashmir
Srinagar, Jammu and Kashmir, India

Venura Herath
Department of Agricultural Biology
Faculty of Agriculture,
University of Peradeniya
Peradeniya, Sri Lanka

ISBN 978-3-030-01414-8 ISBN 978-3-030-01415-5 (eBook)
https://doi.org/10.1007/978-3-030-01415-5

Library of Congress Control Number: 2018960924

This Springer imprint is published by the registered company Springer Nature Switzerland AG
The registered company address is: Gewerbestrasse 11, 6330 Cham, Switzerland

Emeritus Professor Peter Langridge FTSE

Peter was born in Adelaide in 1953 to a Czech mother and a New Zealand father. He was brought up in Canberra where he studied at the Australian National University. When he graduated, he took up a job in Germany at the University of Freiburg. During his 4 years in Germany, he also met and married his German wife, Ursula, who is also a scientist. In 1984, he moved to the University of Adelaide. He became a Professor in 1996 and from 1998 was the inaugural Research Director of the Cooperative Research Centre for Molecular

Plant Breeding. In 2003, Peter became the Chief Executive Officer and Director of the Australian Centre for Plant Functional Genomics (ACPFG) when it was established and remained in this role until 2014. ACPFG was a major research centre based in Adelaide set up by the Australian Federal Government through the Australian Research Council and the Grains Research and Development Corporation. When he left ACPFG, Peter was appointed Emeritus Professor at the University of Adelaide; he is also an Honorary Professor at the Kazakh National Agrarian University. He is a Fellow of the Australian Academy of Technological Sciences and Engineering and an Honorary Fellow of Food Standards Australia and New Zealand (FSANZ) and James Hutton Institute, UK.

Since 2011, Peter has been chair of the Scientific Board of the Wheat Initiative. The Wheat Initiative was established by the G20 group of countries to provide global coordination of wheat research. The secretariat moved from Paris to Berlin at the beginning of 2018. Peter also chairs several science advisory committees for research organisations in Europe and North America. He chaired the steering committee for the CGIAR Research Program on Dryland Cereals and led a major review of biotechnology capabilities across the CGIAR system. In 2011, he chaired an expert scientific panel for the Australian Government on "Food security in a changing world". Peter is Editor-in-Chief of the Journal Agronomy (MDPI Publishers, Switzerland) and associate editor of eight other journals. In 2011, he was selected as the South Australian Scientist of the Year, and he has received other awards in Australia and Europe.

Peter's research has focused on plant molecular biology and the science of plant breeding, and he has published over 300 research papers, books and reviews.

Preface

Human population is increasing at an alarming pace and believed to exceed 9.7 billion by 2050, whereas at the same time the agricultural productivity is decreasing due to the growing environmental constraints as a result of global climate change. Cold stress is one of the widespread abiotic stresses affecting crop productivity particularly in temperate regions. Plants have developed various anatomical, physiological and genetic strategies to cope with the cold stress. Conventional breeding methods have resulted in inadequate success in improving the cold tolerance of vital crop plants through inter-specific or inter-generic hybridization. Therefore, it is of the essence to speed up the efforts for unraveling the biochemical, physiological and molecular mechanisms underlying cold stress tolerance in plants. While quite a few programs have been taken up in leading global research institutes but the pace of development of cold stress tolerant cultivars is not up to the mark when compared to ever-increasing pressure of abiotic stresses including cold stress due to global climate change. Moreover, the intricate genetic mechanisms involved in plant adaptation to cold stresses have been a key obstacle for crop improvement using conventional plant breeding tools. Omics technologies including genomics, transcriptomics and proteomics have facilitated elucidation of complex mechanisms involved in plant adaptation to cold stress. Through this book "Cold Tolerance in Plants - Physiological, Molecular and Genetic Perspectives", we have tried our best to include chapters unfolding the implication of cold stress in plants under climate change scenario and the eventual scientific advancements being applied utilizing the existing high throughput omics technologies to come up with novel strategies to mitigate cold stress by unraveling molecular mechanisms responsible for cold stress in plants.

This book provides systematic and comprehensive reference material for researchers, teachers, and graduate students involved in abiotic stress tolerance studies in plants particularly cold stress using physiological, molecular and genomic tools by unfolding principles and application of recently developed technologies and their application in development of stress resilience in plants against cold stresses. The chapters are written by globally reputed researchers and academicians in the field of plant stress biology. We express sincere thanks and gratefulness to our revered authors, without their untiring efforts this book project would not have been possible. We are also thankful to Springer Nature for providing such opportunity to complete this book project. We are also thankful to all our family members for their support during the entire book project completion.

Srinagar, India Shabir Hussain Wani
Peradeniya, Sri Lanka Venura Herath

Contents

Contributors

Mohammad Arif Ashraf United Graduate School of Agricultural Sciences, Iwate University, Morioka, Japan

Towseef Mohsin Bhat Laboratory of Translational and Evolutionary Genomics, School of Life Sciences Jawaharlal Nehru University, New Delhi, India

Viswanathan Chinnusamy Division of Plant Physiology, ICAR-Indian Agricultural Research Institute, New Delhi, India

Sana Choudhary Genetics Section, Department of Botany, Aligarh Muslim University, Aligarh, India

Muhammad Qudrat Ullah Farooqi Department of Soil Science, School of Agriculture and Environment, Faculty of Science, The University of Western Australia, Perth, WA, Australia

V. C. Dilukshi Fernando Department of Biological Sciences, University of Manitoba, Winnipeg, MB, Canada

Elisabetta Frascaroli Department of Agricultural and Food Sciences (DISTAL), University of Bologna, Bologna, Italy

Venura Herath Department of Agricultural Biology, Faculty of Agriculture, University of Peradeniya, Peradeniya, Sri Lanka

Mohsen Janmohammadi Department of Plant Production and Genetics, Agriculture College, University of Maragheh, Maragheh, Iran

Rohit Joshi Stress Physiology and Molecular Biology Laboratory, School of Life Sciences, Jawaharlal Nehru University, New Delhi, India

Vinay Kumar Department of Biotechnology, Modern College of Arts, Science and Commerce, Savitribai Phule Pune University, Pune, India

Ju Kyong Lee Department of Applied Plant Sciences, College of Agriculture and Life Sciences, Kangwon National University, Chuncheon, Republic of Korea

Valentina Longo National Research Council, Institute for Microelectronics and Microsystems, Lecce, Italy

Jigeesha Mukhopadhyay Post Graduate Department of Biotechnology, St. Xavier's College (Autonomous), Kolkata, West Bengal, India

Abidur Rahman United Graduate School of Agricultural Sciences, Iwate University, Morioka, Japan

Department of Plant Bio Sciences, Faculty of Agriculture, Iwate University, Morioka, Japan

Agri-Innovation Center, Iwate University, Morioka, Japan

Nirala Ramchiary Laboratory of Translational and Evolutionary Genomics, School of Life Sciences Jawaharlal Nehru University, New Delhi, India

Sara Rinalducci Department of Ecological and Biological Sciences (DEB), University of Tuscia, Viterbo, Italy

Aryadeep Roychoudhury Post Graduate Department of Biotechnology, St. Xavier's College (Autonomous), Kolkata, West Bengal, India

Balwant Singh ICAR-National Research Centre on Plant Biotechnology, New Delhi, India

Shabir Hussain Wani Mountain Research Centre for Field Crops, Khudwani, Sher-e-Kashmir University of Agricultural Sciences and Technology of Kashmir, Srinagar, Jammu and Kashmir, India

N. Yahia Genetics and Plant Breeding Laboratory, Department of Biology, Faculty of Sciences of Nature and Life, University of Oran, Oran, Algeria

Zahra Zahra Institute of Environmental Sciences and Engineering, School of Civil and Environmental Engineering, National University of Sciences and Technology, Islamabad, Pakistan

Lello Zolla Department of Science and Technology for Agriculture, Forestry, Nature and Energy (DAFNE), University of Tuscia, Viterbo, Italy

Chapter 1
Cold-Induced Injuries and Signaling Responses in Plants

Jigeesha Mukhopadhyay and Aryadeep Roychoudhury

1.1 Introduction

The phenotypic manifestations of cold injury in plants are highly variable. Both low temperature and rapid fluctuations between heat and cold can severely affect the physiology of plants (Miura and Furumoto 2013). Cold stress inflicts damages to fruit trees, horticultural and landscape plants, as well as crop plants, posing a major threat to sustainable agriculture. Commercially important crop plants have been targeted for stress alleviation in order to increase productivity and yield through interspecific and intergeneric breeding which resulted in limited success; however, transgenic approaches to engineer cold-tolerant plants by manipulation of the key genes of the transcriptional and metabolic cascades have contributed to tolerance mechanisms in affected plants to some extent (Rihan et al. 2017). The present chapter highlights the physiological effects of cold stress and gene regulations on perceiving cold stress signals. Finally, the chapter discusses on the cold tolerance mechanisms, genetic engineering for tolerance, and acclimation that allows adaptation and successful breeding strategies for sustainable growth of plants in the face of cold injuries (Sanghera et al. 2011).

1.2 Cold Injuries: Chilling, Frost, and Freeze

Winter injury as well as freeze and frost injury are often synonymous. Cold injuries, however, are more severely manifested due to extreme temperature fluctuation, rather than prolonged low temperature conditions. Sudden temperature fluctuations

J. Mukhopadhyay · A. Roychoudhury (✉)
Post Graduate Department of Biotechnology, St. Xavier's College (Autonomous), Kolkata, West Bengal, India

S. H. Wani, V. Herath (eds.), *Cold Tolerance in Plants*,
https://doi.org/10.1007/978-3-030-01415-5_1

like rapidly falling temperature and hard freeze can result in stress development and injury in plants that have acquired dormancy, but have not yet fully acclimated (Guy 1990). Acclimation to below-freezing conditions can successfully occur, only if the temperature fall is gradual, whereas deacclimation can occur if extended periods of mild winter occur, and this poses a massive threat to plants if they are suddenly exposed to extremely low temperature conditions. Such deacclimated plants are vulnerable to tissue injury and cold stress (Kalberer et al. 2006). However, prolonged low temperatures, viz., during winter, can also severely damage plants, mainly when the temperature drops below a certain tolerance limit. Plants that are already physiologically weak may be, due to previous stress exposures or due to lack of hardiness and adaptability to the harsh conditions of a specific geographical locale, are more prone to suffer from winter injury (Arora and Rowland 2011). The manifestation of winter injury is highly variable, though buds show maximum susceptibility (Fig. 1.1).

Chilling injury can be defined as damage incurred to plants due to temperature exceeding the freezing point (32°F or 0°C). Maximum susceptibility to chilling injury is shown by plants inhabiting tropical or subtropical climes. Flowers, fruits, and leaves are affected in the sensitive species, and manifestation in the form of purple or reddish wilting leaves is common. Frost and freeze injury are closely related since both lead to membrane damage due to osmotic shock, dehydration stress, and ice crystal formation. Frost damage occurs during radiation freeze,

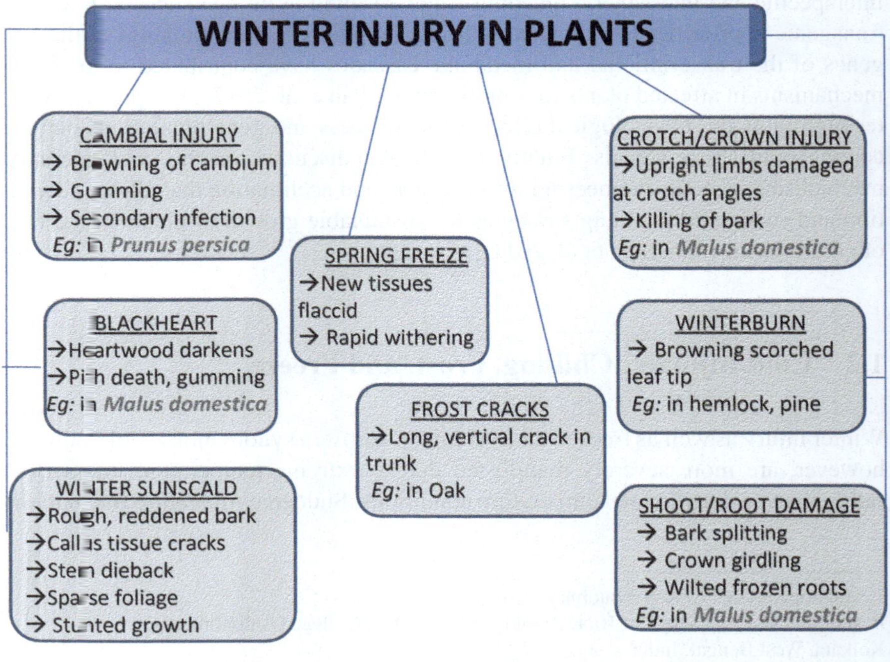

Fig. 1.1 Common manifestations of cold injury in the form of damages to root, shoot, and cambium in chilling-sensitive plants

mainly on calm and clear nights, when plants give off more heat than can be compensated for by the heat received by plants. Thus, it results in a temperature inversion, wherein cold air nearer to the ground is trapped by the warmer air layer above it (air temperature increases with altitude). When the air temperature at the plant level is near or below freezing, the temperature of the plant inevitably is lower than the ambient conditions. Freeze damage, on the other hand, occurs due to advective freezes, when an air mass with below-freezing temperature moves into and occupies an area, displacing warmer air. This causes the temperature of plants low enough to form ice crystals, hence damaging the tissues.

1.2.1 Freezing Injury in Plants

To generalize the term freezing injury, it is mainly concerned with the dysfunctions in physiology of the plant due to freezing of the water contained in plant tissues due to late spring and early fall frosts, low midwinter minima, and rapid temperature fluctuations. Freezing of tissue water is inevitably accompanied by ice formation, which may be intracellular or extracellular (Pearce 2001). Intracellular ice formation may be due to the following: (i) internal nucleation (large polysaccharides or proteins may act as nucleating agents for ice formation) and (ii) penetration (external ice crystals may penetrate into plant cells). Two types of freezing usually occur in plant cells and tissues: (i) vitrification (when rapid freezing of cells to very low temperatures causes the cellular content to get solidified into noncrystalline or amorphous state) and (ii) crystallization (ice crystallization due to gradual drop in temperature may be intracellular or extracellular). The more severe and damaging of the two is intracellular freezing, since it disrupts membrane integrity and can be lethal. Intracellular ice formation in susceptible tender plants is common; however, hardy plants before acclimation may also be affected. Intracellular ice may be formed spontaneously from centers of nucleation in the cytoplasm or may form in cell walls adjacent to intercellular spaces (apoplasm). Sometimes, ice may spread from cell to cell through plasmodesmatal connections. The plasmalemma can serve as a barrier to the entry of ice and hence can partially prevent dehydration, but cells and organelle tend to shrink and succumb to freezing injury to some extent. Thick cuticle can also serve as an effective shield that protects seedlings from external ice. Tissue damages due to freezing injury are characterized primarily by loss of membrane integrity, leakage of metabolites, and perturbations in plasmolysis as well as deplasmolysis.

1.2.1.1 Supercooling and Ice Nucleation

Some "deep supercooled" tissues in hardy plants may also show intracellular freezing. Deep supercooling is a mechanism by which plants avoid freezing injuries. The phenomenon by which water below freezing temperature still maintains its liquid state is known as supercooling. Supercooling can occur in plants when the liquid held in the intercellular spaces does not make the transition from liquid to solid phase, and hence plants can avoid ice crystallization (Wisniewski et al. 2008). Some fruit trees and hardwoods are capable of supercooling down to –35°C; however, below –40°C, ice crystallization is spontaneous. Smaller crystals formed due to rapid freezing usually melt before causing cold injuries. Thus, water can indefinitely remain in supercooled state, unless the temperature falls below this homogeneous ice nucleation temperature or frost and soil ice invade plants through natural openings like stoma, lenticels, and wound sites. If external ice achieves nucleation, it rapidly spreads through vascular tissues, and the number and localization of nucleations depend on the initial extent of supercooling achieved by the plant. If supercooling is sufficient, multiple ice nucleation sites are available for the external ice to intrude.

Deep supercooling is achieved in some woody plants which involves supercooling of an aqueous fraction which is considerably isolated from seedling by an ice layer and is also divided into distinct compartments (Nuener et al. 2010). This compartmentalized pure water spontaneously freezes at –38°C. The presence of solutes depresses the spontaneous nucleation point as seen in lowering of supercooling by experimental addition of solutes to exotherm of shagbark hickory. In xylem parenchyma and flower bud tissues, ice penetration from adjacent frozen tissues is prevented by a barrier formed by undifferentiated cells between floral primordia in bud and nearby frozen stem tissues. Such barriers which prevent propagation of ice into healthy tissues may involve fine microcapillaries of cell wall, in addition to antinucleating chemicals in protoplasm.

1.2.1.2 Mechanism of Injury

(i) Intracellular freezing injury: intracellular freezing is a rapid process which results in flash freezing of cells which then allows ice crystals to propagate throughout protoplast and vacuole. Macromolecular assembly is disturbed due to mechanical tension and dehydration. Membrane integrity is hampered, and cellular compartmentalization is disrupted, resulting in leakage of hydrolyzing enzymes in the affected tissue. (ii) Extracellular freezing injury: extracellular ice imposes desiccation stress on the protoplasm, which is equivalent to drought stress, since water is removed from the cell to the extracellular ice. Dehydration of plant cells due to freezing injury can be lethal, primarily damaging the membrane of frost-injured cells. Following freezing, membrane proteins are rendered insoluble and protein dissociation into subunits occurs resulting in inactivation of enzymes, just as in case of drought stress. Membrane can thus be established as the primary site of desiccation stress. Membrane proteins are denatured due to a number of factors associated with freezing injury like pH imbalance, increased salt concentration, oxidation of sulfhydryl groups, and change in conformation due to water loss.

Tissue shearing in vascular tissues due to ice crystals has been observed in wheat crowns, azalea flower buds, and developing pear fruitlets. Freezing can also lead to shrinkage of protoplasts in injured plants. Lipoprotein membranes show fractures along hydrophobic regions, since intramolecular hydrogen bonds are weakened due to freezing injury.

1.2.2 Chilling Injury in Plants

Chilling injury is the damage incurred to chilling-sensitive plants at temperatures above the freezing point of tissues but lower than 15°C, i.e., injury at low but non-freezing temperature conditions. Plants which show visual manifestations of injury at temperatures exceeding 15°C are referred to as extremely chilling sensitive (Lukatkin et al. 2012). Accordingly, plants can be classified as (i) chilling-sensitive (severely damaged at temperatures above 0°C but below 15°C) and (ii) chilling-resistant (they are able to tolerate low temperature up to a tolerance threshold and show signs of injury only when ice formation occurs).

1.2.2.1 Mechanism of Chilling Injury

The physical phase transition of cellular membranes from flexible liquid crystalline to rigid gel structure at a temperature critical for chilling injury serves as a controlling response. Lowering of temperature in chilling-sensitive species leads to solidification of membrane lipids, which brings about contraction, causing cracks and channels and, consequently, increased permeability. This disturbed regulation of permeability leads to ionic imbalance and ion leakage from tissues. Enzyme activity is also hampered, since suitable temperature condition for optimum activity is not available. The temperature-induced phase change of membrane lipids is reversible till degenerative damage has been caused to the plant (Parkin et al. 1989).

1.3 Alterations in Cell Membrane: Marker for Chilling Stress Injury

The phase transition of cellular membranes from flexible, fluid state to rigidified solid state serves as a marker for detecting chilling-induced injury in plants. Such phase transition is characterized by the appearance of gel-like sites or microdomains in the plane of the lipid bilayer, which are partially or completely protein-free. Multiple membrane changes are detectable in stressed chilling-sensitive plants, viz., decrease in membrane elasticity, reduced compliance, preventing the inclusion of lipids in membrane composition, reduction of fluidity and hence flexibility of membrane lipids, and inactivation of membrane-bound enzymes, including H^+-ATPase with increased lateral diffusion of phospholipids, sterols, and proteins in the plasma membrane (Kasamo et al. 1992; Kasamo and Noushi 1987).

Membrane functioning under chilling stress is dependent on the membrane lipids (Routaboul et al. 2000). In chilling-sensitive plants, membrane integrity is affected due to chilling-induced degradation of galactolipids and phospholipids, which result in an increased pool of free fatty acids. In stressed plants, a distinct change in molar ratio of sterols is observed, and increase in ratio of sterols/phospholipids resulted in decreased membrane fluidity on lowering of temperature (Whitaker 1993). A marked increase in unsaturated fatty acids, phospholipid accumulation in tissues, and depletion in sterols and sterol esters are physiological manifestations of chilling stress in sensitive species (Kojima et al. 1998; Kaniuga et al. 1999).

Membrane transport is severely affected due to reduced permeability associated with increased viscosity in response to low temperature. Hence, water uptake and sugar translocation were reduced in chilling-sensitive species. Distinct changes are also observed in the membrane proteins exposed to chilling stress (Fig. 1.2). Protein conformation is lost, and the nonprotein components of enzymes are released, resulting in changes in the allosteric control of activity and kinetic parameters. Low temperature-induced enzyme inactivation is also mediated by protein-lipid interactions in the membrane. Molecular ordering of membrane lipids changes due to low

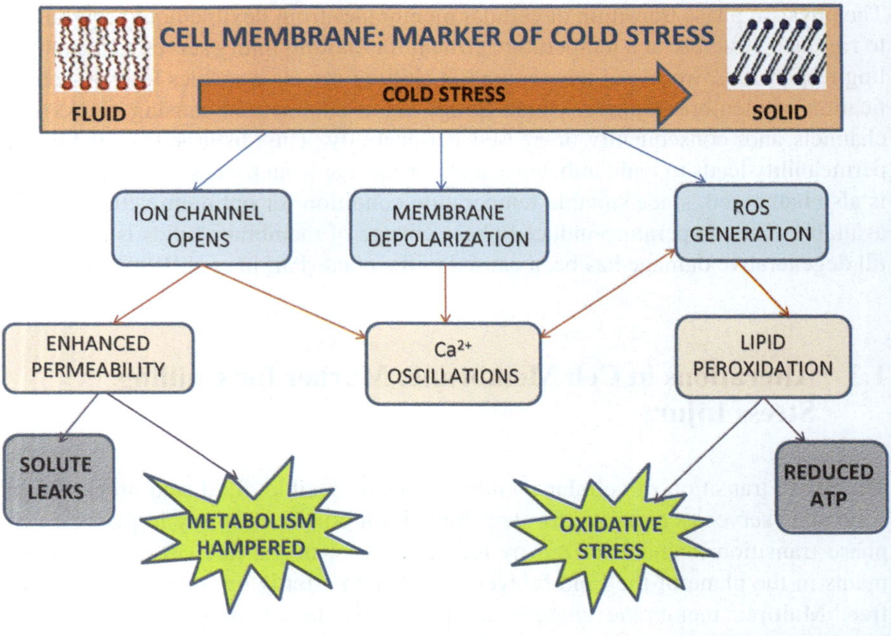

Fig. 1.2 Scheme for initiation of physiological changes on perception of cold stress cues with the plasma membrane serving as the main marker for detection of chilling stress; ROS (reactive oxygen species), ATP (adenosine triphosphate)

temperature exposures. These changes are accompanied by lowered ATP levels and increased membrane permeability. Hence, the membrane is the seat for the detection of chilling-induced injury in plants. Prolonged exposure to chilling stress disrupts membrane integrity and compartmentalization, solute leakage, and increase in the activation energy barrier for membrane-bound enzymes, thus jeopardizing the overall physiological status of the affected species.

1.4 Cold Perception and Downstream Signaling

Environmental cues perceived by the plant result in an intricate network of downstream signaling cascades. Different receptors at the cellular level are involved in receiving the external signals and, in turn, transfer them intracellularly. Plants are sensitive to both magnitude and rate of temperature fluctuations. Thermal responses in plants in the face of cold stress involve a complex intracellular machinery and genetic regulation. There are two principal transcriptional pathways that are activated in response to cold stress, C-repeat (CRT)/dehydration responsive element (DRE)-binding factor (CBF/DREB)-dependent and CBF/DREB-independent. The transcription factor, CBF, acts as a master regulatory player and is induced by the binding of trans-acting factors to the promoter regions of the *CBF* gene (Fowler and Thomashow 2002). The constitutively expressed ICE1 (Inducer of CBF Expression 1) binds to the corresponding cis element on the *CBF* promoter and elicits the ICE1-CBF cold-responsive pathway, which is conserved in diverse plant species (Chinnusamy et al. 2003).

1.4.1 Stress Perception Through Plasma Membrane Rigidification

The physiological responses of plants to stress are variable; however, membrane rigidification is a common response, since rapid fall in temperatures induces membrane to become rigid at microdomains. Signaling pathways involving calcium waves have also been worked out in alfalfa and *Brassica napus*, where cold stress induction led to actin cytoskeletal rearrangement and loss of fluidity of the plasma membrane, activation of Ca^{2+} channels, and, hence, rapid calcium oscillations (Orvar et al. 2000). Increased cytosolic Ca^{2+} levels induce the expression of *cold-responsive (COR)* genes, which can be activated artificially by a membrane rigidifier like dimethyl sulfoxide (DMSO) even at 25°C, while its expression is inhibited by a membrane fluidizer like benzyl alcohol even at 0°C (Sangwan et al. 2001). Ca^{2+} is the ubiquitous second messenger and is a major player in the cold-responsive signaling pathways (Knight et al. 1996), and mechanosensitive calcium channels have been found to be involved in cold acclimation. Intracellular calcium ion channels implicated in *COR* expression are activated by cyclic ADP-ribose and inositol-1, 4, 5-triphosphate (IP_3). A typical Ca^{2+}-responsive signaling pathway consists of Ca^{2+}-activated phospholipase C and D (PLC, PLD) which produce IP_3 and phosphatidic acid, respectively, and, in turn, activate IP_3-gated calcium channels.

Rise in intracellular calcium levels can be perceived by calcium-dependent protein kinases (CDPKs) and calmodulins (CAMs) and salt overly-sensitive 3-like (SOS3-like) or calcineurin B-like (CBL) proteins. That CDPKs play a functional role in cold stress signaling was proved through a transient expression system in maize leaf protoplasts where a constitutively active form of an *Arabidopsis* CDPK (CDPK1) activated the expression of abscisic acid (ABA)-responsive promoter of *HVA1* gene (Sheen 1998). Hence, CDPKs were proved to have a positive role in mediating cold signaling; however, CAMs, CBLs, and SOS3-like proteins are negative regulators of such signaling cascades.

1.4.2 The CBF-COR Regulon: Transcriptional Machinery

Cold stress response is mediated by a gene regulatory network in which the CBFs are critical transcription factors, as they are involved in the control of the *COLD-REGULATED (COR)* genes through the CBF-COR regulon (Thomashow 1999). CBFs are also involved in the drought and salinity stress-responsive pathways, thereby proving that there exists an intricate cross-talk mechanism between the different forms of abiotic stress. The CBFs belong to the APETALA/ethylene response element-binding protein (AP2/EREBP) transcription factor family (Stockinger et al. 1997) and are modulated by upstream regulators like inducer of CBF expression 1 (ICE1), *high expression of osmotically responsive 1* (*HOS1*) gene, and MYB15 (Agarwal et al. 2006). ICE1 is a constitutively expressed myc-like bHLH (basic helix-loop-helix) transcription factor, which binds to *CBF3* gene promoter, inducing its expression, and is degraded via the ubiquitin-proteasomal pathway through the cold signaling attenuator HOS1, an E3 ubiquitin ligase (Dong et al. 2006).

The promoter regions of *COR* genes consist of one or multiple copies of the C-repeat/DRE with the highly conserved CCGAC core sequence. The CBFs or DREBs control ABA-independent expression of *COR* genes in response to cold stress, which indicates that ABA may be able to potentiate cold-induced CBF signaling, but ABA and cold stimuli may not be concurrent. CBF-DREB1 is involved in transcriptional response to cold as well as osmotic stress-regulated genes, whereas CBF/DREB2 is exclusively responsive to cold stress, and not to salinity or osmotic stress conditions, and is controlled by ICE1 transcription factor. This further provides an insight into the cross-talk of abiotic stress-responsive signaling pathways. Microarray analysis of CBF-overexpressing transgenic plants identified several CBF target genes involved in signaling, transcription, osmolyte biosynthesis, reactive oxygen species (ROS) detoxification, membrane transport, hormone metabolism, and stress response and can sufficiently induce cold tolerance in diverse plant species, e.g., *AtCBF1* of tomato enhanced oxidative stress tolerance under chilling stress as well as enhanced tolerance to dehydration stress, whereas *AtCBF3* of rice resulted in enhanced tolerance to drought and high salt, together with a marginal increase in chilling tolerance.

The coordination of regulation of cold tolerance and plant development is regulated by signaling hormones like ABA, gibberellic acid (GA), and auxin (Lee et al. 2010). During cold stress, growth retardation appears to be regulated by CBFs through DELLA proteins, which are localized in the nucleus and represses growth in *Arabidopsis*, whereas GA-stimulated degradation of DELLA leads to promotion of growth enhanced by CBFs (Achard et al. 2008). ICE1, which regulates *CBF* expression, is constitutively expressed; however, only exposure to cold temperature leads to ICE1-induced expression of *CBF* and other cold stress-responsive genes. Overexpression of HOS1, a negative regulator of cold-responsive pathways, leads to substantial decline in ICE1 protein and its target genes, since HOS1 targets ICE1 to proteasomal degradation machinery via the ubiquitin pathway, and this leads to hypersensitivity to cold stress. The role of ICE1 in regulating photosynthesis and transpiration through stomatal development has also been investigated, and this establishes a bridge between stomatal development and responsiveness to environmental stress signals. ICE1 physically interacts and dimerizes with bHLH transcription factors like SPEECHLESS (SPCH), MUTE, and FAMA, thus regulating stomatal development (Kanaoka et al. 2008). ICE1 also negatively regulates expression of MYB15, which itself is a negative player in cold stress-responsive pathways in *Arabidopsis*. Overexpression of *MYB15* leads to reduced expression of *CBFs*, whereas *myb15* T-DNA knockout mutants showed enhanced cold induction of *CBFs* (Agarwal et al. 2006). Thus, regulation of CBF expression is dependent on the interaction between the positive and negative regulators of cold stress-responsive pathways.

Calmodulin-binding transcription activator (CAMTA) family proteins have also been studied as transcriptional regulators of CBF2 expression, which is proved through mutational studies involving *camta3* mutants, which, under cold stress, showed reduced levels of CBF2 in comparison to the wild-type plants. Further, double mutants of *camta/camta3* exhibit hypersensitivity to freezing stress in comparison to wild-type plants, establishing the role of CAMTA proteins in activating *CBF2* expression (Doherty et al. 2009).

1.4.3 CBF-Independent Regulons

Genetic screening led to the identification of transcription factors that mediate cold-responsive pathways and, subsequently, are responsible for conferring cold tolerance in *Arabidopsis*. These transcription factors include high expression of HOS9, a homeodomain protein, and HOS10, an R2R3-type MYB (myc-like basic helix-loop-helix), and transcriptome analysis showed distinct CBF and HOS9 regulons (Fig. 1.3). HOS10 acts through an ABA-dependent cold-responsive pathway, since it is responsible for the activation of ABA biosynthesis gene *NCED3* (9-cis-epoxycarotenoid dioxygenase) leading to ABA accumulation during cold stress (Zhu et al. 2004).

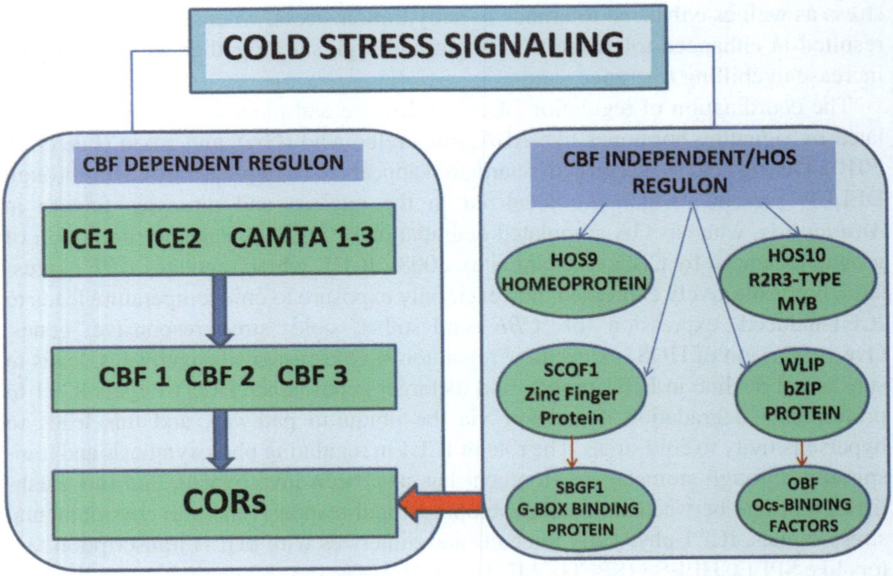

Fig. 1.3 The cold stress-responsive transcriptome showing control through CBF-dependent and CBF-independent regulons; CBF C-repeat/drought-responsive element-binding factor, ICE inducer of CBF expression, CAMTA calmodulin-binding transcription activator, COR cold responsive, HOS high expression of osmotically responsive genes, MYB myc-like beta helix-loop-helix, SCOF soybean cold-inducible factor, OBF ocs element-binding factor, WLIP wheat low temperature-induced protein

In response to freezing, the expression of *COR* genes in *Arabidopsis* transgenics was upregulated due to constitutive overexpression of soybean C2H2-type zinc finger protein, soybean cold activation factor (SCOF1), which, in turn, interacts with the conserved soybean G-box-binding factor 1 (SGBF1) (Kim et al. 2001). Another example of overlapping abiotic stress-responsive pathways is given by wheat low temperature-induced protein 19 (WLIP19), a basic leucine zipper protein (bZIP), which is induced by cold, drought, and ABA. WPLIP19 is involved in the activation of expression of *COR* genes in wheat by physically interacting and dimerizing with bZIP transcription factor TaOBF1in wheat (ocs element-binding factor 1 in *Triticum aestivum*). These observations altogether present examples of CBF-independent regulons involved in cold stress response.

1.5 Metabolic Changes in Cold Stress-Injured Plants

In cold-sensitive species, exposed to low temperature conditions, cytoplasmic gelling occurs due to increase in viscosity of the cytoplasm. This leads to impediment of biochemical reactions in the membrane, hence affecting the metabolic health of the plant. Prolonged exposure to cold stress leads to metabolic disorders, and normal physiological processes of the plant are disturbed. One of the probable causes of metabolic changes is the uncoupling between energy obtained during respiration and its effective utilization. Chilling injury may also result in accumulation of toxic end products due to metabolic imbalance, which ensues after the temperature rises to normalcy, indicating that this is a secondary dysfunction associated with heating under post chilling conditions. Cold-induced metabolic changes can be simulated by ectopically expressing *CBF* genes under warm conditions, thereby proving that the CBF-dependent regulon has a key role in reconfiguration of the low temperature metabolome. This is consistent with the results from *Arabidopsis*, where the strain (Cape Verde Islands-1 ecotype) having a weak regulon and being deficient in CBF-regulated metabolites was incapable of cold acclimation (Cook et al. 2004).

The soluble sugars act as important signaling molecules during cold stress (Rolland et al. 2006). Compartmentalization of sugars is an important aspect of plant response to cold stress. In cereals exposed to freezing stress, fructans stored in vacuole are mobilized to fructose and then exported to the intracellular liquid, where it interferes with the adhesion between wet plant surface and extracellular ice. Hence, extracellular ice cannot invade into the living cell. In *Arabidopsis*, differential distribution of respiratory metabolites takes place during cold acclimation (Talts et al. 2004). A slight decrease in lipid hydrolysis has been reported in chilling-sensitive fruit tree like papaya, with concomitant increase in soluble solids. Cold-injured roots of sweet potato are unable to synthesize carotenoid pigments, and accelerated depletion of ascorbic acid has been reported in not only sweet potato but also cold-injured banana and pineapple. Dark coloration of fruit pulp, which is symptomatic of cold injury, has been observed in chilled banana due to accumulation of tannin and its oxidized products, as well as elevated levels of dihydroxyphenylalanine (DOPA) and tyrosine, whose polymerization and oxidation products result in dark pigmentation. Chilled sweet potato roots also showed a marked increase in chlorogenic acid. Similarly, pepper seeds also had higher levels of chlorogenic acid and total polyphenols.

Protein metabolome is also severely affected under cold stress. An active aspartate shuttle has been evidenced to play an important role in cold acclimation, in which pools of amino acids like asparagine, glutamine, and glutamate shift from the plastid to cytosol (Hoermiller et al. 2016). The experiments involved two distinct mutants in primary carbon metabolism (*pgm*, the starchless mutant of the plastidial phosphor glucomutase, and *spsa1*, a mutant defective in the dominant sucrose phosphate synthase), which demonstrate that inhibition of primary carbon metabolism limits the ability of the plants to respond to cold stress and also alter the patterns of intracellular allocation of metabolites in response to an imposed stress. The protein metabolome undergoes dramatic changes during cold stress response, since stress induces accumulation of soluble proteins to facilitate acclimation and tolerance. The level of high molecular weight glycoproteins also increase, as has been reported in mulberry bark cells and orchard grass tissue. Cold-induced cryoprotective proteins have been isolated from spinach chloroplast, and their nuclei also contained elevated levels of high molecular weight proteins at low temperature conditions, whereas low molecular weight proteins are predominantly present under warmer conditions (Guy 1990). Cold stress response and acclimation leads to the activation of genes that can impart tolerance, and these genes are usually not expressed under normal conditions. Thus, synthesis of proteins that contribute to acclimation is increased in response to cold stress, and this was established through the isolation of translation competent polysomes from cold-acclimated tissues as well as through reported increase in rRNA and RNA polymerase activity during cold stress (Sarhan et al. 1997). Housekeeping proteins continue to be synthesized under low temperature conditions to maintain basal level of cellular metabolism. The ability to synthesize proteins faster in acclimated tissues is attributed to increased rRNA levels and protein synthesis capacity. However, it is to be noted that the set of proteins synthesized in response to low temperature stress is not as highly conserved as heat shock proteins, pointing toward the diversity of cold stress conditions and variability in stress response (Sakai and Larcher 1987). The changes in protein metabolism inevitably include altered activity of enzymes, especially those that are associated with the membrane. High molecular weight proteins tend to replace their smaller, low molecular weight counterparts during cold stress response, since they are more efficient considering kinetic parameters and K_m values and hence can impart tolerance. One such example is the invertase enzyme from wheat which shifts from lower to higher molecular weight forms during acclimation to low temperature. Increase in enzyme levels coincided with the increase in soluble protein content under low temperature conditions. Increase in activity of enzymes associated with respiratory pathways, such as glucose-6-phosphate dehydrogenase, lactate and isocitrate dehydrogenase are characteristic of cold-induced regulation of enzymes. Isozymic and conformational alterations in enzymes under cold stress exposure render them cryostable. In winter, new isozymic variants were reported for ATPases, esterases, acid phosphatases, leucine aminopeptidases, and peroxidases, and this winter-specific enzyme machinery contributed to freeze stability under adverse winter conditions.

Supramolecular interactions are also changed on exposure to low temperature, as evidenced by the oligomerization of light-harvesting chlorophyll *a/b* protein of Photosystem II and chlorophyll *a*-protein complex of Photosystem I in rye thylakoids. Studying the lipid profile revealed that there is a specific decrease in levels of *trans*-Δ^3-hexadecenoic acid of phosphatidyldiacylglycerol in thylakoid membranes due to the cold-induced oligomerization of light-harvesting complex (Krupa et al. 1987).

The cold-induced expression of distinct *GST* genes encoding glutathione S-transferase superfamily proteins, which contribute to cold tolerance, has been investigated in *Brassica oleracea* (Vijayakumar et al. 2016). Secondary metabolites also tend to accumulate in response to cold stress, along with elevated levels of sugar (Livingston et al. 2009), polyamines (Kovacs et al. 2010), anthocyanins (Christie et al. 1994), and glucosinolates (Baskar et al. 2012). In cold-injured tissues, activity of enzymes like glutathione S-transferase (GST), glutathione reductase (GR), superoxide dismutase (SOD), and glutathione peroxidase (GPX) increased manifold, along with elevated levels of nonenzymatic antioxidants like tripeptide thiols and vitamins (Janska et al. 2010), signifying that all these enzyme proteins are functionally important for cold tolerance.

1.6 Cytophysiological Changes Due to Cold Injury

Cell membrane damage and loss of compartmentalization are primary markers for cold-induced injury in plants. Besides cell membrane vesicularization, rupture of plasmalemma, destruction of endoplasmic reticulum, and alterations in Golgi body are also reported as a consequence of cold stress. Ultrastructure of mitochondria is also affected, leading to swelling and degeneration, matrix enlightenment, shortening of cristae, and decrease in their number, inevitably affecting oxidative phosphorylation (Yin et al. 2010). Prolamellar plastids were improperly formed, causing swelling and structural changes in chloroplast membranes. Thus, mitochondrial and chloroplast damage hampers both the photosynthetic and respiratory machineries. In the next sections, we probe into the physiological effects of cold injury and stress-induced alteration in photosynthetic efficiency and respiration, which eventually weakens the cold-susceptible plants physiologically.

1.6.1 Photosynthesis

The decline in the rate of photosynthesis, associated with cold stress, is related to fall in temperature and prolonged exposure to low temperature, and this effect persists long after the chilling-sensitive plants have been transferred to warmer conditions post chilling (Liu et al. 2004; Strauss et al. 2006). The loss of photosynthetic efficiency in cold-injured plants can be attributed to inhibition of solute (carbohydrate) transport from leaves via phloem; stomatal limitation; damaged photosynthetic machinery, especially the light-harvesting complex associated with Photosystem I (PSI); inhibition of electron transport; uncoupling of electron transfer and energy conservation; and finally loss of activity of key enzymes of the Calvin cycle and C_4-pathway. Light chilling (chilling of sensitive plants in presence of light) has more damaging impact than chilling in dark since the photosynthetic apparatus is affected mainly due to photoinhibition and photooxidation as a result of excess excitation energy sequestered from the photosynthetic machinery. Photoinhibition or decline in photosynthetic efficiency under excessive illumination and chilling is directly proportional with light intensity and fall in temperature.

Photooxidative damage incurred to photosystems housed in the chloroplast membranes is manifested by increased lipid peroxidation and degeneration of photosynthetic pigments like chlorophyll, carotene, and xanthophylls.

The effects on photosynthesis in cold-injured *Hibiscus* plants have been investigated (Paredes and Quiles 2015). The study involved chilling stressed illuminated plants of *Hibiscus rosa-sinensis*, and the involvement of chlororespiratory enzymes and ferredoxin-mediated cyclic electron flow was dissected in these cold-susceptible plants, where cold stress resulted in reduction of efficiency of photosynthesis and subsequently, electron transport. This study revealed the role of cyclic electron flow in protecting the photosystems during cold stress, as established by the increased activity of electron donation by NADPH and ferredoxin to plastoquinone and elevated levels of PGR5 polypeptide which is an essential component of cyclic electron flow around PSI. Cold stress response also involved increase in levels of chlororespiratory enzymes NDH (NADPH dehydrogenase) complex and PTOX (plastid terminal oxidase). The study of photosynthetic potential in cold-stressed *Rhododendron chrysanthemum* (Zhou et al. 2017) revealed that cold stress could lead to a significant reduction in electron transport rate of Photosystem II (PSII), accompanied with an increase in excitation pressure (1-qP, where qP refers to the photochemical quenching). Photochemical efficiency of PSII was also affected in Mediterranean *Citrus albidus* L. and *Quercus ilex* L. (Oliveira and Penuelas 2005).

1.6.2 Mitochondrial Damage and Respiration

Respiratory rate decline during chilling stress is correlated with mitochondrial damage and lowering of kinetic energy and inhibition of respiratory enzymes. Decreased respiration coupled with increased utilization of energy-rich phosphates at chilling temperatures leads to decline in ATP levels. It has also been observed that cold stress shifts respiration to alternate pathways from the cytochrome path of electron transport in seedlings (Ribas-Carbo et al. 2000), since these accessory pathways functionally contribute to cold acclimation by reducing superoxide and reactive oxygen species (ROS) generation in mitochondria (Hu et al. 2010).

The mitochondrial respiratory chain in *Arabidopsis* has been studied under cold stress (Golzalez et al. 2007) which revealed a coordinated, tissue and developmentally dependent response of mitochondrial complex components encoded by the nuclear genome. Chilling leads to pronounced breakdown of mitochondrial proteins, mimicking protein degradation under drought stress, and often, modified or damaged versions of mitochondrial proteins (oxidized or S-nitrosylated forms) are produced abundantly under cold stress (Taylor et al. 2011). Proline is a key player in temperature stress recovery, where it serves as nitrogen and energy source, as well as a compatible osmolyte and effective ROS scavenger. Proline turnover takes place in the mitochondria where it is imported and converted into glutamate via the Pro/P5C cycle by proline dehydrogenase (PDH) and $^{1}\Delta$-pyrroline-5-carboxylate dehydrogenase (P5CDH); however, under temperature stress, the cycle may occur reciprocally as established by the findings that free proline accumulation in cold-treated cauliflower (*Brassica oleracea* var. *botrytis*) and PDH mutant plants impart frost resistance under cold stress (Hadi et al. 2011). Morphological changes in mitochondria under cold stress has been studied in chilled mung bean (*Vigna radiata*) suspension cells and cucumber root tips where extensive mitochondrial swelling during cold recovery was accompanied by appearance of internal translucence and vesicular structures (Lee et al. 2002). Aberrant ring-shaped mitochondria in chilled *Arabidopsis* mesophyll cells (Vella et al. 2012) and disorganized bursting mitochondria in recovering *Episcia reptans* (Murphy and Wilson 1981) were reported in association with chilling injury. Respiratory declines due to oxidative damage in mitochondria coincide with upregulation of rescue pathways mediated by alternative oxidase (AOX) (Sugie et al. 2006), suggesting importance of AOX in cold acclimation (Fiorani et al. 2005; Armstrong et al. 2008). The redox homeostasis under cold stress is maintained by cold-responsive uncoupling protein 1 (UCP1) in *Arabidopsis* which facilitates photosynthesis under stress conditions by smooth running of the electron transport chain (Sweetlove et al. 2006), while other uncoupling proteins like plant uncoupling mitochondrial proteins 1, 4, and 5 (PUMP 1, 4, and 5) and cold shock proteins like CSP310 of wheat also play important roles (Kolesnichenko et al. 2002).

Accumulation of thermostable dehydrins in response to freezing and cold stress has also been reported in mitochondria of cereal crops like wheat and rye (Borovskii et al. 2002), whose function is to stabilize mitochondrial membrane and matrix proteins. Respiratory rate decline could also be associated with reduced capacity for respiratory NADH oxidation under cold stress, as has been investigated in leaves of *Solanum tuberosum* L., cv. Desiree (Svensson et al. 2002). This decrease is accompanied by reduced levels of NDA proteins (*nda* genes encode non-proton pumping respiratory chain NADH dehydrogenases) and internal rotenone-insensitive NADH oxidation, as has been observed in mitochondria isolated from cold-treated plants. These proteins are major players in flexible tuning of redox homeostasis in cytosol and matrix, in response to ATP demand; hence, reduced expression of *nda* genes during cold stress could possibly lead to respiratory energy dissipation. The ROS generated in response to cold stress lead to activation of antioxidant system in plants as has been discussed earlier. In order to scavenge ROS and protect plants from oxidative damage, plants have evolved efficient antioxidant systems which can be categorized as antioxidant enzymes; lipid-soluble, membrane-associated antioxidants like α-tocopherol, β-carotene, and ubiquinone; and water-soluble antioxidants like reduced glutathione (GSH) and ascorbate via ascorbate-glutathione cycle. Antioxidant response to cold stress has been thoroughly investigated in oil plants *Jatropha curcas* and *Jatropha macrocarpa* (Spano et al. 2017), in which antioxidant enzymes like ascorbate peroxidase, glutathione peroxidase, guaiacol peroxidase, and catalase contributed significantly to ROS scavenging and cold acclimation during cold-induced oxidative damages.

1.7 Interplay of Phytohormones during Cold Stress

Hormonal interplay during chilling stress involves intricate cross-talk between signaling pathways via the major phytohormones that play functionally distinct roles in mediating cold-responsive signaling. This signaling network integrates external stress cues into endogenous developmental programs, thus initiating an appropriate response for alleviation of the stress.

1.7.1 Gibberellin (GA)

GA is such a phytohormone whose signaling is regulated centrally by DELLA which is GRAS protein, whose five members have been characterized in *Arabidopsis*, viz., GIBBERELIC ACID INSENSITIVE (GAI), REPRESSOR OF GAI (RGA), and RGA-LIKE 1, 2, and 3 (RGL1, RGL2, and RGL3). Cold-induced expression of GA 2-oxidases leads to increased hydroxylation and inactivation of bioactive GA which could be correlated to suppressed growth and delayed flowering due to stress conditions. *CBF1* overexpression during cold stress response enhanced DELLA accumulation through transcriptional regulation of RGL3 (Achard et al. 2008). Furthermore, to establish that absence of GA and promotion of freezing tolerance coincide, mutational analysis was carried out, and it was shown that *Arabidopsis gai* mutant (a constitutive signaling mutant) was tolerant to freezing, whereas DELLA knockout lines *gai-t6* and *rga-24* were hypersensitive to freezing.

1.7.2 Brassinosteroid (BR)

BRs are plant polyhydroxysteroids which play an essential role in various developmental and physiological processes such as stem elongation, root growth, vascular differentiation, leaf epinasty, and reproduction (Fujioka and Yokota 2003; Kim and Wang 2010; Sasse 2003). The role of BRs in alleviating chilling stress has been extensively investigated in chilling-sensitive crops such as maize and cucumber (*Cucumis sativus*). That BRs functionally contribute to freezing tolerance is indicated by enhanced expression of *CBFs* and *CBF* target *COR47* in *Arabidopsis* upon exogenous application of BR following chilling, which successfully imparted cold tolerance. Regulation of antioxidant enzymes during cold storage by BR has also established the functional significance of BR in imparting cold stress tolerance in Washington Navel Orange (*Citrus sinensis* L.) (Ghorbani and Pakkish 2014).

1.7.3 Cytokinin (CK)

CKs are adenine derivatives with isoprenoid or aromatic side chains that control directional growth processes like gravitropism and are also involved in control of both CBF-dependent and CBF-independent regulons during cold stress response. *CYTOKININ RESPONSE FACTOR (CRF2)* and *CRF3* encoding APETALA2 transcription factors regulate *Arabidopsis* lateral root (LR) initiation when subjected to low temperature conditions (Jeon et al. 2016). Similarly, *CRF4* is cold inducible and imparts freezing tolerance as has been investigated in *Arabidopsis*. Analysis of the cold transcriptome reveals that cytokinin and a subset of two component signaling system are involved in cold stress signaling and response which comprise histidine kinases, phosphotransfer proteins, and response regulators.

1.7.4 Abscisic Acid (ABA)

Abscisic acid is an isoprenoid hormone which has been christened the "universal stress hormone" since it is a key player in the response pathways to a plethora of abiotic stress conditions, besides its role in mediating seed dormancy and abscission (Nakashima et al. 2009). Its role in cold stress signaling has been investigated, particularly with respect to its regulation through the CBF-dependent pathway. A cold- and ABA-inducible transcription factor, MYB96, has been studied in *Arabidopsis* (Lee et al. 2015) which reveals that this transcription factor is involved in *CBF* induction and cold acclimation, which was compromised in *myb96* knockout lines, whereas cold-responsive pathways were upregulated by MYB96 overexpression. Moreover, MYB96 indirectly interacts with cold-responsive transcription factors which regulate *CBF* expression by binding to HEPTAHELICAL PROTEIN (HHP), HHP1, HHP2, and HHP3, which physically interact with ICE1, CAMTA, and ICE2, respectively (Chen et al. 2010).

Another regulator of *CBF* expression is the cold-inducible transcription factor Open Stomata 1 (OST1), a serine/threonine kinase, which is activated by ABA, and its role in cold stress response has been established by hypersensitivity to freezing in *ost1* knockout lines, whereas increased freezing resistance was achieved through overexpression. Hence, ABA ubiquitously participates in abiotic stress signaling and has a major role in stress tolerance in stress-injured plants.

1.7.5 Ethylene

The gaseous phytohormone, ethylene regulates many aspects of plant life cycle such as seed germination, leaf senescence, fruit ripening, and abscission, and its role in cold stress response has also been studied. However, the level of ethylene in response to cold stress was found to be variable from one species to the other, as evidenced by increased levels (through enhanced expression of ethylene biosynthetic genes) in cold-stressed *Arabidopsis* (Catala et al. 2014), in contrast to rapid decline in cold-treated *Medicago truncatula* (through reduction of ethylene precursor 1-aminocyclopropane-1-carboxylic acid or ACC) (Zhao et al. 2014). The role of ethylene as a negative regulator of cold-responsive signaling is indicated by studies in *M. truncatula* where administering ethylene biosynthesis inhibitor 2-aminoethoxyvinyl glycine (AVG) enhanced freezing tolerance post cold treatment, whereas hypersensitivity to freezing was reported on treatment with ethylene releaser ethephon, correlated with a compromised induction of expression of *MtCBF1*, *MtCBF3*, and *MtCAS15*, which belong to *COLD ACCLIMATION SPECIFIC* (*CAS*) gene family equivalent to *COR* genes (Zhao et al. 2014). However, the positive role of ethylene in *Arabidopsis* freezing tolerance disputes this exclusivity of negative role of ethylene, as proved through mutational studies where ACS octuple mutant with extremely low levels of ethylene was hypersensitive to freezing (Catala et al. 2014). Another study on grapevine reported the cold-induced enhanced release of ethylene through modulation of expression of ETHYLENE RESPONSE FACTOR (ETR) 057, and the evidence was supported by enhanced cold tolerance on exogenous ACC application and reduced resistance to cold on treatment with aminoethoxyvinylglycine (AVG). Hence, though the role of ethylene in cold response remains controversial, it can be considered an important candidate regulator of cold stress signaling in susceptible plants.

1.7.6 Auxin

Biologically active auxins are indole acetic acid (IAA) and indole butyric acid (IBA) which essentially control all aspects of plant development, from embryogenesis to senescence, and are key players in intricate hormonal cross-talks influencing various developmental stages. The role of auxin in cold stress response is not clearly dissected; however, cold stress-mediated interruption of auxin transport has been established through analysis of auxin response and molecular mechanisms in *Arabidopsis*, where auxin signaling mutants *axr1* and *tir showed* reduced gravity response but normal response to cold treatment (equivalent to wild type). This indicated that stress affects auxin transport, without altering signaling (Shibasaki et al. 2009). Moreover, exogenous application of auxin analogues on canola (*Brassica napus*) stimulated accumulation of cryoprotective metabolites and soluble sugars, aiding in acclimation. Auxin influx and efflux carriers that determine spatial gradients of auxin concentration also influence auxin-regulated directional gravitropic growth. Such gravitropic growth is temperature-sensitive, as proved by its inhibition in *Arabidopsis* due to cold, correlated with altered distribution patterns of auxin in roots and repressed basipetal auxin transport caused by impaired intracellular cycling of auxin efflux carriers PIN2 and PIN3 (Shibasaki et al. 2009). Another interesting observation came from the study of SIZ1, a central regulatory component of cold signaling pathway which stabilizes ICE1 under cold conditions, by repressing its polyubiquitination (Miura et al. 2007), and negatively regulates phosphate starvation-induced root architecture remodeling through control of auxin patterning (Miura et al. 2011).

1.7.7 Melatonin

Melatonin (N-acetyl-5-methoxytryptamine), besides being an animal neurohormone, is also a regulator of plant developmental and stress-responsive pathways. The role of melatonin in alleviating cold stress has been investigated in Bermuda grass (*Cynodon dactylon* L. Pers.), where exogenous application of melatonin prevented cold damage to Photosystem II through activation of multiple antioxidants and accumulation of freeze-protective secondary metabolites, amino acids, organic acids, soluble sugars, and sugar alcohols (Fan et al. 2015), besides maintaining membrane stability, improving process of Photosystem II, and inducing cold-responsive metabolic changes. Oxidative damage in wheat was also alleviated by application of melatonin to seedlings that enhanced the activities of superoxide dismutase, guaiacol peroxidase and ascorbate peroxidase, and other ROS scavengers. Cold stress tolerance by melatonin treatment in *Arabidopsis* is attributed to melatonin-mediated upregulation of *CBFs* and DREBs, as well as cold-responsive gene *COR15a*, transcriptional activator *CAMTA1*, and antioxidant genes *ZAT10* and *ZAT12*, following cold stress (Bajwa et al. 2013).

1.7.8 Salicylic Acid (SA)

SA is a phenolic compound, which is predominantly involved in triggering defense against biotic stress factors and pathogen hypersensitive response (HR). However, its role in abiotic stress signaling has also been studied, and it has been reported that SA accumulation and cold-induced growth suppression could be correlated, since SA-overaccumulating *constitutive expression of PR* (*cpr1*) mutant plants showed retarded growth under cold conditions, whereas SA deficiency led to cell cycle progression (G1/S phase transition) through enhanced transcription of cytokinin-regulated D-type cyclin CYCD3 in *Arabidopsis*. In chilling sensitive cucumber, SA was proposed to activate *CBF* expression, as established by the application of SA biosynthesis inhibitor, L-α-aminooxy-β-phenylpropionic acid (AOPP), which suppressed the expression of *CBFs* and *COR47* in cold-treated seedlings (Dong et al. 2014). SA-induced chilling and oxidative stress tolerance can also be correlated to gibberellin homeostasis, as studied in cold-stored tomato, through C-repeat/DREB factor pathway and antioxidant enzyme system (Ding et al. 2015). Foliar application SA also influenced the redox status of canola (*B. napus* L.) through activation of antioxidant enzymes during cold stress exposure.

1.7.9 Jasmonic Acid (JA)

JA is an oxylipin whose levels increase under cold stress, as has been reported in different plant species like rice, wheat, and *Arabidopsis* through increased expression of JA biosynthetic genes and repression of genes encoding JA catabolic enzymes. Exogenous application of JA imparts cold tolerance in *Arabidopsis* through enhanced *CBF* expression, whereas compromise in JA biosynthesis leads to hypersensitivity to freezing (Hu et al. 2013). Altered freezing tolerance is mediated by JASMONATE ZIM-DOMAIN (JAZ) proteins, JAZ1 and JAZ4, which physically interact with ICE1 and ICE2, and JA-mediated cold-induced growth inhibition occurred through stabilization of DELLA proteins (Yang et al. 2012).

1.8 Freezing Stress Tolerance: Natural Mechanism and Engineering

1.8.1 Natural Mechanism: Dormancy and Vernalization

1.8.1.1 Dormancy and Cold Stress

Endodormancy is a distinct physiological process that is sensitive to, but does not need, cold or other external conditions to be induced (Faust et al. 1991), i.e., it is controlled by both internal factors and seasonal fluctuations in temperature and photoperiod. Hence, cold stress and its consequent acclimation can be closely associated with endodormancy. Endodormancy can result either in absence of cold or due to intermittent cold conditions, whereas continuous above freezing temperature results in endodormancy release. During chilling conditions of late autumn and early winter, potential accumulation of FT (Flowering Locus T) results (Lin et al. 2007) which influences the timing of endodormancy release and bud flush. FT is a mobile protein that travels from leaves to shoot apices, and hence, the formation of callose plugs in autumn inhibits their translocation. Such plugs are removed during freezing winter temperature through the activation of gibberellic acid-responsive genes that regulate GH17 proteins which break down callose, and hence FT translocation and accumulation can ensue (Rinne et al. 2001). Hence, endodormancy can be considered as one of the survival strategies of cold stressed plants in the temperate regions, and its key regulator *FT* has evolved from lineage-specific duplications. Furthermore, the functional role of bud set timing synchronized with endodormancy release is being investigated to determine how plants respond to selective environmental pressures, by combining genomics and functional and developmental approaches and correlating ecological mechanisms and phenotypic adaptations to adverse environmental conditions (Rohde and Bhalerao 2007).

1.8.1.2 Vernalization, Flowering, and Freezing Tolerance

When an extended period of cold makes plants competent for flowering, it is referred to as vernalization. Unlike endodormancy, shoot apices of vernalized plants maintain a basal level of mitotic division for supporting the vegetative condition. In *A. thaliana*, there is evidence of epigenetic control of vernalization responsiveness. The flowering repressor gene *FLC* and its five *MADS AFFECTING FLOWERING* (*MAF*) paralogs are epigenetically silenced by the Plant HomeoDomain-Polycomb Repressive Complex 2 (PHD-PRC2), which initiates trimethylation of histone 2 lysine 27 (H3K27me3) (Ratcliffe et al. 2003). This complex becomes progressively localized to the first intron of *FLC* during cold exposure (Shindo et al. 2005; Angel et al. 2011; Strange et al. 2011) and initiates two noncoding transcripts from this intron [COOL-ASSISTED INTRONIC NONCODING RNA (COLDAIR)] and 3′-UTR (COOLAIR) which negatively regulate *FLC* transcription (Heo and Sung 2011). This consequently results in the *FT*-regulated change from vegetative to inflorescence identity, induction of MADS-box genes such as *FRUITFULL (FUL)* and *APETALA1 (AP1)*, floral development, formation of siliques, and subsequently seeds.

Vernalization in crop species, wheat and barley (Pooideae subfamily), has also been investigated where the responsiveness is guided by the three loci, *VERNALIZATION1 (VRN1)*, *VRN2*, and *VRN3*. *VRN1* is homologous to the *Arabidopsis* floral development genes *AP1*, *CAULIFLOWER (CAL)*, and *FUL*, and its expression is under epigenetic control through chromatin remodeling. Transcription factors interacting with the cis elements of the promoter regions of the *VRN1* gene mediate chromatin modifications that repress its expression prior to winter, whereas its long-day induction is prevented by the zinc finger *CO*-like gene *VRN2* (Alonso-Peral et al. 2011). Vernalization can be considered an accelerating factor for floral development and shortening of the juvenile or vegetative phase in plants. In winter cereals, the timing of vernalization saturation, deacclimation, downregulation of cold-induced genes, and reduced reacclimation potential are regulated by the *VRN1* locus, which indicate that the probability of freezing damage after warm season depends both on vernalization and photoperiod. This subsequently led to the investigation of the role of vernalization and the *VRN1* gene on freezing tolerance (Ergon et al. 2016). Thus, we can conclude that freezing tolerance, vernalization, and the onset of flowering are interlinked, and the genetic regulation shows overlapping activities of the key players in manipulating the plants to undergo flowering by overriding the stress cues.

1.8.1.3 Engineering Cold Stress Tolerance Through Transgenic Approach

In transgenic approach, several key regulatory players of cold stress-responsive pathways can be selected as candidates for manipulation and upregulation of cold-tolerant genes that can alleviate the damaging effects of low temperature injury in plants (Wani et al. 2008). The *CBF* genes are the key regulatory elements in cold-responsive signaling pathways and hence serve as potential targets of genetic manipulation to engineer stress tolerant plants. Transgenic *Arabidopsis* plants overexpressing *CBF1* showed freezing tolerance while avoiding the negative impact of cold stress on development and growth characteristics. Such overexpression also activates *COR* homologous genes in *Arabidopsis*. Constitutive overexpression of cold inducible transcription factors like CBF1 has been shown to impart cold stress tolerance, through introduction of *CBF1* cDNA into chilling-sensitive tomato (*Solanum lycopersicum*) under the control of strong CaMV35S promoter (Hsieh et al. 2002). Another candidate target is the *CBF4*, a close *CBF/DREB*1 homolog, whose overexpression alleviated both freezing and drought stress in *Arabidopsis*.

Arabidopsis DREB1A overexpression played the dual role of alleviation of damaging impact of both drought and low temperature stress in tobacco, and regulation of transgene expression through stress-inducible *RD29A* promoter minimized the cold-induced inhibition of plant growth (Kasuga et al. 2004). *DREB1A* gene and its two homologs were induced by low temperature stress, whereas expression of *DREB2A* and its homolog was observed in response to drought stress cues. Hence, DREBs present examples of integration of stress-responsive pathways involving both cold and dehydration cues.

CBFs are negatively regulated by an upstream transcription factor *MYB15* (an *R2R3-MYB* family protein) in *Arabidopsis* (Agarwal et al. 2006) which recognizes cis elements in the promoter of *CBF* called *MYB* recognition elements (*MYBRS*). Since it is a negative regulator of *CBF*, a transgenic approach involving *MYB* mutation resulted in enhanced cold acclimation by efficient expression of *CBFs* during cold stress, whereas *MYB* overexpression reduced freezing tolerance through negative regulation of *CBF*. The CBF regulon also provides the *COR* target for transgenic modification. *COR* gene expression and freezing tolerance in transgenics have been established through studies in which activation of *COR* by a zinc finger protein product of *SCOF-1* enhanced tolerance in soybean (Kim et al. 2001). On the other hand, the CBF-independent regulon presents the example of a protein eskimol (*ESK1* gene product) which, when mutated, resulted in constitutive cold tolerance (Xin and Browse 1998). Generating cold-tolerant transgenic plants through introduction of cold-shock proteins (CSPs) like CspA from *Escherichia coli* and CspB from *Bacillus subtilis* has successfully promoted stress adaptation in multiple species (Castiglioni et al. 2008).

Structural genes sometimes serve as targets for transgenic manipulation if they contribute to stress alleviation and tolerance. Genetically engineered tobacco plants overexpressing chloroplast glycerol-3-phosphate acyltransferase (*GPAT*) gene (which participates in desaturation of phosohatidyl glycerol fatty acid) from squash and *Arabidopsis* accumulate unsaturated fatty acids in the cell wall which enhanced tolerance to cold stress (Murata et al. 1992). Cold-induced expression of gene *TPP* (*trehalose-6-phosphate phosphatase*) is followed by accumulation of trehalose in rice tissues exposed to cold stress (Pramanik and Imai 2005). Hence, cold tolerance has been shown to be achieved through overexpression of *TPS* (*trehalose-6-phosphate synthase*) and *TPP* which enhanced trehalose accumulation in transgenic rice and tobacco. Transgenic *Arabidopsis* expressing plant phosphatase *AtPP2CA* enhanced cold acclimation response during freezing. Constitutive expression of kinase *NPK1* also improved tolerance to chilling as well as to other forms of abiotic stress (Kovtun et al. 2000). Overexpression of rice mitogen-activated protein kinase *OsMAPK5* conferred tolerance to a plethora of abiotic stress conditions including freezing stress, as observed in rice seedlings (Xiong and Yang 2003). Stress-responsive genes encoding calcineurin B-like protein-interacting protein kinases *OsCIPK03* and *OsCIPK12* are also involved in stress tolerance response in rice (Xiang et al. 2007). Calcium-dependent protein kinases are major players in different signal transduction pathways, and its cold-induced expression in rice leads to accumulation of the gene product OsCDPK7 and OsCDPK13, which serve as positive regulators of cold tolerance pathways (Wan et al. 2007). The *Late Embryogenesis Abundant* (*LEA*) genes also act as suitable targets for manipulation for their functional importance in abiotic stress response. Expression of the citrus gene-encoding *LEA* protein *CuCOR19* enhanced cold tolerance in transgenic tobacco (Hara et al. 2003), whereas introduction of wheat dehydrin gene *WCOR410* in strawberry imparted freezing tolerance (Houde et al. 2004). All these transcriptomic analyses and transgenic attempts established the significance of CBF-DREB-mediated cold stress responses.

1.8.1.4 Integration of Polyamines in Cold Tolerance Mechanisms

Intracellular accumulation of endogenous polyamines (PA) occurs in response to cold stress as they contribute to plant response to low temperature conditions. The increase in levels of diamine putrescine (Put) has been reported in cold-stressed *Arabidopsis* (Kaplan et al. 2004). The role of Put is to modulate ABA biosynthesis at the transcriptional level and hence regulate cold-induced, ABA-dependent responsive pathways. PA pre-treatment to seedlings of wheat showed alleviation of chilling injuries, where priming with spermidine (Spd) decreased the chilling stress-induced lipid peroxidation and membrane leakage, supported by endogenous increase in PA under low temperature conditions. The main function of PA is to maintain the catalytic activity of antioxidant enzymes and hence prevent oxidative damages associated with cold injury. Polycationic molecules also stabilize cellular membranes and minimize changes in membrane permeability, hence preventing loss of fluid. They also play a role in regulating hydrogen peroxide production, indirectly modulating plant defense mechanisms during chilling stress. Since high cellular levels of PA can be correlated to stress alleviation and tolerance, obtaining plants with elevated PA levels by genetically manipulating their biosynthesis can provide improved tolerance to abiotic stress conditions, including low temperature. For this, usually PA biosynthetic genes or those involved in the maintenance of PA homeostasis are targeted. *Arabidopsis* plants transformed with *SPDS* (*spermidine synthase*) cDNA from *Cucurbita ficifolia* under the control of strong, constitutive promoter, CaMV35S led to freezing tolerance (Kasukabe et al. 2004). *SPDS* overexpressor lines also exhibited enhanced activity of other PAs like spermine (Spm) and Put. In the PA biosynthetic pathway, overexpression studies have been conducted on cold-induced genes *ADC1* and *ADC2* (*arginine decarboxylase* 1 and 2) which promoted freezing tolerance in transgenic *Arabidopsis* and tobacco (Alcazar et al. 2005). Besides *ADC*, enhanced transcript levels of *SAMDC* (*S-adenosylmethionine decarboxylase*) have been reported in cold stress response. Transformation of tobacco plants with carnation *SAMDC* cDNA under strong promoter CaMV35S led to accumulation of Spd, Spm, and Put improving tolerance to multiple abiotic stress conditions like salinity and acidic and oxidative stress, other than freezing stress (Wi et al. 2006). The increased titers of Put on overexpression of *SAMDC* were actually the result of high Spd accumulation which was actively interconverted to Put by acetylation. PAs are also involved in integration of signaling pathways modulating ROS- (hydrogen peroxide), NO- (nitric oxide), and ABA-mediated responses. PA levels are controlled by ABA, and this has been established through studies that showed impairment of stress-induced expression of *ADC2*, *SPDS1*, and *SPMS* (*spermine synthase*) genes in ABA-deficient and ABA-insensitive *Arabidopsis* mutants. This is, however, an example of an ABA-dependent cold-responsive pathway which is regulated by a bZIP (basic leucine zipper) family of transcription factors that interact with ABRE motifs, and this control is distinct from the CBF/DREB1 regulon which is ABA-independent.

1.9 Epigenetic Regulation of Chilling Responses

Low temperature-induced epigenetic modifications on perception of cold stress cues involve posttranslational histone modifications, DNA methylation, histone variant incorporation, and chromatin remodeling. Epigenetic modifications persist between generations through adaptive transgenerational plasticity and hence inherited by the progenies, through mitotic cell divisions, and transmitted to the next generation. Chromatin perception of ambient temperature has been investigated through study of *arp6* mutants of *Arabidopsis* (*ARP6* gene encodes a protein which is a subunit of the SWR1 chromatin-remodeling complex) (Kumar and Wigge 2010). SWI/SNF-type ATP-dependent chromatin remodeling complexes are evolutionarily conserved multiprotein machineries which regulate accessibility of DNA and chromatin structure to transcription factories. An ATPase component of this complex, BRM (BRAHMA), is a key target of ABA-dependent dephosphorylation through PP2C phosphatases, thereby indicating its role in ABA-dependent abiotic stress-responsive pathways. The presence of histone H2A.Z variant in the nucleosomes maintains the promoters in the repressed state till the perception of an activation signal (Li et al. 2015), i.e., the promoters of quiescent genes remain on standby mode till transcription initiation. Temperature-mediated changes in nucleosome composition and accessibility of nucleosomal DNA to promoter sequences have been analyzed by studying the H2A.Z occupancy at the *HSP70* promoter (HSPs or heat shock proteins maintain cells and protects proteins from denaturation during temperature stress) in response to varying ambient temperature which demonstrates that nucleosomes containing H2A.Z variant respond to varying ambient temperature conditions and in turn coordinate the transcriptome. Alterations in chromatin dynamics also mark cold stress responses in *Cannabis sativa* L., where the induction of *COR* genes is mediated by changes in chromatin structure (Mayer et al. 2015). In response to cold acclimation treatment, there was a marked initial increase in global DNA methylation that could be reverted during treatment only in cold-tolerant varieties. Significant increase in methylcytosine levels was also observed at the *COR* gene loci, on deacclimation which suggested an epigenetic mechanism of cold acclimation through locus-specific DNA methylation. Further, acetylation or deacetylation of H3K9 plays an important role in transcription regulation of *COR* genes (Pavangadkar et al. 2010), and its significance is proved through mutational studies in *Arabidopsis* where mutants for histone deacetylases develop a hypersensitivity to freezing temperatures. Plants with defect in histone acetyltransferases are incapable of optimally inducing *COR* genes during cold acclimation (Vlaconasios et al. 2003). In cereal crops like rice (*Oryza sativa*), differential acetylation of Histone H3 at the regulatory region of *DREB1b* (*Drought Response Element Binding 1b*) promoter facilitates chromatin remodeling and transcription activation in response to cold stress (Roy et al. 2014). The rice ortholog of DREB1, *OsDREB1b*, is transcriptionally induced by cold stress; hence targeting this gene for overexpression confers increased tolerance to salinity and cold stress. On perception of cold stress cues, there is significant change in nucleosome arrangement at the upstream region of *OsDREB1b*, cis elements, and TATA box at the core promoter. Significant increase in acetylation levels of H3K14 and H3K27, hyperacetylation of H3K9, and enrichment of RNA Pol II occupancy at the promoter region lead to activation of transcription machinery through epigenetic mechanism. Hence, alterations of chromatin conformation lead to upregulation of the cold stress-responsive gene, increasing tolerance, and alleviation of stress.

1.10 Epigenetic Control of Vernalization Responses

Vernalization promoted flowering through epigenetic repression of flowering suppressor *FLC* in *Arabidopsis*, and such modifications are transmitted to the progeny (Jablonka and Raz 2009). There is also evidence of epigenetic memory of vernalization in winter cereals, which results from alterations in histone H3 lysine methylation levels throughout the extent of the *VRN1* gene. The increased H3K9 trimethylation and decreased H3K27 trimethylation are indicative of a vernalization-induced active chromatin state within *VRN1* gene (vernalization response is mediated by the stable induction of *VRN1* gene promoter). *VRN1* downregulates the floral repressor *VRN2* while allowing induction of the floral activator *FT* or *VRN3* which accelerates subsequent changes leading to floral development in cold-stressed plants. Vernalization and cold tolerance may also be mediated by DNA demethylation which results from the action of histone demethylases (Liu and Secombe 2015). Regulation of another key gene of the vernalization pathway is *FLC* (*FLOWERING LOCUS C*), which encodes a MADS-box transcription factor that represses genes involved in floral initiation, and hence the gene product is accumulated prior to cold adaptation by vernalization. This gene is also epigenetically controlled and remains transcriptionally silenced when temperature conditions are favorable, allowing expression of the downstream genes involved in floral initiation, such as *SOC1* (*SUPPRESSOR OF CONSTANS 1*) and *FT* (*FLOWERING LOCUS T*). Inhibition of *FLC* expression is correlated with elevated levels of repressive histone modifications through methylation of histone H3 lysine 9 and 27, as well as loss of histone modifications associated with active transcription such as H3 acetylation and H3 lysine 4 methylation (Bastow et al. 2004; Finnegan and Dennis 2007; Schmitz et al. 2008). The cellular memory of transcriptional repression of *FLC* is maintained through successive cell divisions by mitotic inheritance of repressive histone modifications of the gene (Sung et al. 2006).

1.11 Role of microRNAs in Chilling Stress Response

MicroRNAs (miRNAs) are a class of small, noncoding RNAs that regulate target genes of abiotic stress-responsive pathways. The broad classification of small regulatory RNAs includes small interfering RNAs (siRNAs) along with miRNAs, which play a crucial role in plant development, and its functionality has been reported at the early phase of anther development when the susceptibility to cold stress is maximum. Small RNA (smRNA) and trans-acting small interfering RNA (tasiRNA) have been studied in thermosensitive genic male sterile (TGMS) lines of wheat (*Triticum aestivum*) which regulate auxin signaling pathway in relation to developmental response to cold stress (Tang et al. 2012). TGMS lines are hypersensitive to low temperature during meiosis stage. Under stress conditions, miRNAs guide anther development by mediating cold-induced abnormal activity of target *ARFs* (*Auxin Responsive Factors*), required for anther development under cold conditions.

Stress-induced miRNAs target repressors of stress-responsive pathways as well as positive regulators of signaling pathways which are inhibited by stress conditions. Accordingly, under stress conditions, there is an over- or under-expression of certain miRNAs which contributes to stress tolerance. Both miRNAs and siRNAs are loaded into AGO (ARGONAUTE) protein containing RISC (RNA-Induced Silencing Complex) which guides transcriptional or posttranscriptional target regulation through RNA-directed DNA methylation (RdDM), a pathway also involved in transgenerational inheritance of epigenetic stress markers and hence maintenance of stress memory. Cold stress results in the downregulation of *MET1*, leading to demethylation of mobile genetic elements (transposons) in *Zea mays* (Shan et al. 2013), whereas in *Arabidopsis* seedlings, subjection to cold stress leads to activation of retrotransposons. The siRNA pathway plays a key role in inhibition of retrotransposition in response to abiotic stress cues.

The miRNAs are involved in intricate overlapping stress regulatory networks (Khraiwesh et al. 2012). A specific example is presented by miR172 which plays the dual role of mRNA cleavage, as well as repression of translation (Jones-Rhoades et al. 2006). Elevated expression of this cold-inducible miRNA leads to repression of translation of *Arabidopsis* flowering development gene product AP2 (APETALA 2), resulting in early flowering and defects in floral organ identity (Chen et al. 2005; Axtell et al. 2006) during cold stress response. Another example is presented by cold-inducible miR169, which inhibits the expression of *XTH* (xyloglucan endotransglucosylase/hydrolase) genes, which function in cell elongation through increase in cell wall elasticity in response to stress cues.

Chloroplast is a major sensor of environmental stress cues, and tolerance during abiotic stress conditions is achieved through stabilization of lamellar membrane systems of chloroplasts through alterations in lipid composition. The gh-miR397a-2, which targets chlorophyll a-binding protein P4, is upregulated in low temperature stress in cotton (*Gossypium hirsutum*) (Wang et al. 2016). Further, the role of miRNAs has been investigated in alleviation of oxidative stress, and miR398 has been implicated in targeting the mRNA of chloroplastic copper/zinc superoxide dismutase (CSD). The reduced expression level of miR398 correlated with the elevated CSD expression leads to scavenging of superoxide radicals and hence detoxification during oxidation stress.

The seedling tolerance to cold stress is also dependent on auxin signaling and modulation of auxin response through miR160g in cotton which targets ARFs (Auxin Responsive Factors) that bind Auxin Responsive Promoters (ARPs), thereby regulating the expression of auxin-responsive genes. Hence, downregulation of miR160g releases their repressing effect on ARFs during cold stress; otherwise, attenuation of plant growth and development may occur under stress conditions. Regulation of a specific miRNA miR319 during cold stress has been studied in sugarcane (Thiebaut et al. 2012). Small RNA transcriptomes have also been investigated to provide insights into miRNA-mediated wheat inflorescence development under cold stress by targeting floral development genes such as ARF, SPB (Squamosa Promoter Binding like Protein), MADS-box (MCM1, AG, DEFA, SRF), MYB, SPX (SYG1, Pho81, XPR1), TCP (TEOSINTE BRANCHED, Cycloidea and PCF), and PPR (Pentatricopeptide repeat) (Zhang et al. 2017). Hence, analysis of small RNA transcriptomes and their target genes provide a novel insight into abiotic stress tolerance mediated by miRNA which serve as an additional layer of gene regulation in cold stress response and hence increase productivity of commercially important crops.

1.12 Conclusion

The transcriptomic analysis of cold-tolerant plants has provided a thorough insight into the intricate genetic regulatory networks that lead to diverging cold-responsive signaling pathways during freezing/chilling stress. Such regulons have been investigated in order to identify suitable target genes that are chief players in mediating stress response and hence can be considered candidates for genetic manipulation for establishment of stable, tolerant transgenics of commercially important plants. Priming with polyamines is another approach through which stress alleviation has been possible; hence engineering pathways for enhancing polyamine levels endogenously other than external application and priming are novel tolerance conferring approaches. Physiologically, the photosynthetic and respiratory machinery are the most affected due to cold stress; thus such machineries have also been elaborately analyzed in order to engineer approaches to prevent chloroplast and mitochondrial damage and improve the scavenging of free radicals and reactive oxygen species through efficient antioxidant systems and maintenance of redox homeostasis in the face of cold stress. Membrane studies have also been conducted since they are the primary markers of cold stress damage through osmotic imbalance and lipid peroxidation, along with alterations in metabolism which have been studied extensively. The overlapping genetic networks between drought and cold stress-responsive pathways provide insights into cross-talks between the different forms of abiotic stress; there are reported examples of many genes that participate in both pathways, hence serving as targets for manipulation. The interplay of phytohormones also provides a comprehensive view into the biochemical changes involved in cold stress response, and alteration of such hormone levels can regulate tolerance. Thus, integrating the current established understanding of the low temperature transcriptome, proteome, and metabolome, it is now possible to come up with approaches involving transgenics as well as epigenetic control so that the cytophysiological damage to cold stressed plants can be prevented. The novel techniques for imparting tolerance through microRNAs have opened up an alternative field for stress alleviation and tolerance engineering.

1.13 Future Perspectives

The complexity of the mechanisms of plant adaptation to stress arises from the multiple players involved in the elaborate stress-responsive transcriptome. An integrated understanding of the multiple aspects of cold-responsive pathways, the gene regulatory networks, the sensory mechanisms, and changes at the cytological, physiological, as well as metabolic levels will help in designing experimental approaches that can alleviate the damaging impact of cold stress on important fruit yielding, ornamental, as well as crop plants. The regulation of transcriptional response to stress, though has been deciphered to some extent, there still lies immense scope in probing further into the transcriptome and identifying novel genes of yet unidentified function in stress-responsive signaling pathways. The proteome and metabolome also need to be investigated extensively since accumulation of relatively newly characterized compatible solutes like glycinebetaine and trehalose indicates that genes encoding for enzymes synthesizing such metabolites can be novel targets of transgenic manipulation. Epigenetic regulations and their study in stress tolerance are in its nascent state and have to be investigated and researched on, in future, globally, in order to engineer novel techniques to confer cold stress tolerance in susceptible plants, exposed to freezing conditions due to their geographical locations as well as for adaptation to changing environment and climate. For this, microRNA approach has already been worked upon and has shown some promise in targeting various loci for strengthening plant tolerance to abiotic stress conditions. Thus, besides the utilization of regulon biotechnology, to dissect the CBF-COR dependent and independent regulons by research groups worldwide, the metabolome and epigenome also need to be extensively studied for a successful global and integrated research effort at alleviating cold stress.

Acknowledgments Financial assistance from the Council of Scientific and Industrial Research (CSIR), Government of India, through the research grant [38(1387)/14/EMR-II] to Dr. Aryadeep Roychoudhury is gratefully acknowledged.

References

Achard P, Renou JP, Berthomé R, Harberd NP, Genschik P (2008) Plant DELLAs restrain growth and promote survival of adversity by reducing the levels of reactive oxygen species. Curr Biol 18:656–660

Agarwal M, Hao Y, Kapoor A, Dong CH, Fujii H, Zheng X, Zhu JK (2006) A R2R3 type MYB transcription factor is involved in the cold regulation of CBF genes and in acquired freezing tolerance. J Biol Chem 281:37636–37645

Alcázar R, Garcia-Martinez JL, Cuevas JC, Tiburcio AF, Altabella T (2005) Overexpression of ADC2 in *Arabidopsis* induces dwarfism and late-flowering through GA deficiency. Plant J 43:425–436

Alonso-Peral MM, Oliver SN, Casao MC, Greenup AA, Trevaskis B (2011) The promoter of the cereal *VERNALIZATION1* gene is sufficient for transcriptional induction by prolonged cold. PLoS One 6:e29456

Angel A, Song J, Dean C, Howard M (2011) A Polycomb-based switch underlying quantitative epigenetic memory. Nature 476:105–108

Armstrong AF, Badger MR, Day DA, Barthet MM, Smith PMC, Millar AH, Whelan J, Atkin OK (2008) Dynamic changes in the mitochondrial electron transport chain underpinning cold acclimation of leaf respiration. Plant Cell Environ 31:1156–1169

Arora R, Rowland LJ (2011) Physiological research on winter hardiness: deacclimation resistance, reacclimation ability, photoprotection strategies, and a cold acclimation protocol design. Hortscience 46:1070–1078

Axtell MJ, Jan C, Rajagopalan R, Bartel DP (2006) A two-hit trigger for siRNA biogenesis in plants. Cell 127:565–577

Bajwa VS, Shukla MR, Sherif SM, Murch SJ, Saxena PK (2013) Role of melatonin in alleviating cold stress in *Arabidopsis thaliana*. J Pineal Res 56:238–245

Baskar V, Gururani MA, Yu JW, Park SW (2012) Engineering glucosinolates in plants: current knowledge and potential uses. Appl Biochem Biotechnol 168:1694–1717

Bastow R, Mylne JS, Lister C, Lippman Z, Martienssen RA, Dean C (2004) Vernalization requires epigenetic silencing of FLC by histone methylation. Nature 427:164–167

Borovskii GB, Stupnikova IV, Antipina AI, Vladimirova SV, Voinikov VK (2002) Accumulation of dehydrin-like proteins in the mitochondria of cereals in response to cold, freezing, drought and ABA treatment. BMC Plant Biol 2:5

Catala R, Lopez-Cobollo R, Castellano MM, Angosto T, Alonso JM, Ecker JR, Salinas J (2014) The *Arabidopsis* 14-3-3 protein RARE COLD INDUCIBLE 1A links low-temperature response and ethylene biosynthesis to regulate freezing tolerance and cold acclimation. Plant Cell 26:3326–3342

Castiglioni P, Warner D, Bensen RJ, Anstrom DC, Harrison J, Stoecker M, Abad M, Kumar G, Salvador S, D'Ordine R, Navarro S, Back S, Fernandes M, Targolli J, Dasgupta S, Bonin C, Luethy MH, Heard JE (2008) Bacterial RNA chaperones confer abiotic stress tolerance in plants and improved grain yield in maize under water-limited conditions. Plant Phyisol 147:446–455

Chen CC, Liang CS, Kao AL, Yang CC (2010) HHP1, a novel signaling component in the cross-talk between the cold and osmotic signaling pathways in *Arabidopsis*. J Exp Bot 61:3305–3320

Chen CC, Simard MJ, Tabara H, Brownell DR, McCollough JA, Mello CC (2005) A member of the polymerase beta nucleotidyltransferase superfamily is required for RNA interference in *C. elegans*. Curr Biol 15:378–383

Chinnusamy V, Ohta M, Kanrar S, Lee BH, Hong X, Agarwal M, Zhu JK (2003) ICE1: a regulator of cold-induced transcriptome and freezing tolerance in *Arabidopsis*. Genes Dev 17:1043–1054

Christie PJ, Alfenito MR, Walbot V (1994) Impact of low-temperature stress on general phenylpropanoid and anthocyanin pathways: enhancement of transcript abundance and anthocyanin pigmentation in maize seedlings. Planta 194:541–549

Cook D, Fowler S, Fiehn O, Thomashow MF (2004) A prominent role for the CBF cold response pathway in configuring the low-temperature metabolome of *Arabidopsis*. Proc Natl Acad Sci U S A 101:15243–15248

Ding Y, Sheng JP, Li SY, Nie Y, Zhao JH, Zhu Z, Wang ZD, Tang XM (2015) The role of gibberellins in the mitigation of chilling injury in cherry tomato (*Solanum lycopersicum* L.) fruit. Postharvest Biol Technol 101:88–95

Doherty CJ, Van Buskirk HA, Myers SJ, Thomashow MF (2009) Roles for *Arabidopsis* CAMTA transcription factors in cold-regulated gene expression and freezing tolerance. Plant Cell 21:972–984

Dong CH, Agarwal M, Zhang Y, Xie Q, Zhu JK (2006) The negative regulator of plant cold responses, HOS1, is a RING E3 ligase that mediates the ubiquitination and degradation of ICE1. Proc Natl Acad Sci U S A 103:8281–8286

Dong CJ, Li L, Shang QM, Liu XY, Zhang ZG (2014) Endogenous salicylic acid accumulation is required for chilling tolerance in cucumber (*Cucumis sativus* L.) seedlings. Planta 240:687–700

Ergon A, Melby TI, Rognli O (2016) Vernalization requirement and the chromosomal *VRN1*-region can affect freezing tolerance and expression of cold-regulated genes in *Festuca pratensis*. Front Plant Sci 7:207

Fan J, Hu Z, Xie Y, Chan Z, Chen K, Amombo E, Chen L, Fu J (2015) Alleviation of cold damage to photosystem II and metabolisms by melatonin in bermudagrass. Front Plant Sci 6:925

Faust M, Liu D, Millard MM, Stutte GW (1991) Bound versus free water in dormant apple buds – a theory for dormancy. Hortscience 26:887–890

Finnegan EJ, Dennis ES (2007) Vernalization-induced trimethylation of histone H3 lysine 27 at FLC is not maintained in mitotically quiescent cells. Curr Biol 17:1978–1983

Fiorani F, Umbach AL, Siedow JN (2005) The alternative oxidase of plant mitochondria is involved in the acclimation of shoot growth at low temperature. A study of *Arabidopsis* AOX1a transgenic plants. Plant Physiol 139:1795–1805

Fowler S, Thomashow MF (2002) *Arabidopsis* transcriptome profiling indicates that multiple regulatory pathways are activated during cold acclimation in addition to the CBF cold response pathway. Plant Cell 14:1675–1690

Fujioka S, Yokota T (2003) Biosynthesis and metabolism of brassinosteroids. Annu Rev Plant Biol 54:1371–1464

Ghorbani B, Pakkish Z (2014) Brassinosteroid enhances cold stress tolerance of Washington navel orange (*Citrus sinensis* L.) fruit by regulating antioxidant enzymes during storage. Agric Conspec Sci 79(2):109–114

Gonzalez DH, Welchen E, Attallah CV, Comelli RN, Mufarrege EF (2007) Transcriptional coordination of the biogenesis of the oxidative phosphorylation machinery in plants. Plant J 51:105–116

Guy CL (1990) Cold acclimation and freezing tolerance: role of protein metabolism. Ann Rev Plant Physiol Plant Mol Bio 41:187–223

Hadi F, Gilpin M, Fuller MP (2011) Identification and expression analysis of CBF/DREB1and COR15 genes in mutants of *Brassica oleracea* var. botrytis with enhanced proline production and frost resistance. Plant Physiol Biochem 49:1323–1332

Hara M, Terashima S, Fukaya T, Kuboi T (2003) Enhancement of cold tolerance and inhibition of lipid peroxidation by citrus dehydrin in transgenic tobacco. Planta 217:290–298

Heo JB, Sung S (2011) Vernalization-mediated epigenetic silencing by a long intronic noncoding RNA. Science 331:76–79

Hoermiller I, Naegele T, Augustin H, Stutz S, Weckwerth W, Heyer A (2016) Subcellular reprogramming of metabolism during cold acclimation in *Arabidopsis thaliana*. Plant Cell Environ 40:599–560

Houde MH, Dallaire S, N'Dong D, Sarhan F (2004) Overexpression of the acidic dehydrin WCOR410 improves freezing tolerance in transgenic strawberry leaves. Plant Biotechnol J 2:381–387

Hsieh TH, Lee JT, Charng YY, Chan MT (2002) Tomato plants ectopically expressing *Arabidopsis* CBF1 show enhanced resistance to water deficit stress. Plant Phyisol 130:618–626

Hu WH, Xiao YA, Zeng JJ, Hu XH (2010) Photosynthesis, respiration and antioxidant enzymes in pepper leaves under drought and heat stresses. Biol Plant 54:761–765

Hu Y, Jiang L, Wang F, Yu D (2013) Jasmonate regulates the inducer of CBF expression-C-repeat binding factor/DRE binding factor1 cascade and freezing tolerance in *Arabidopsis*. Plant Cell 25:2907–2924

Jablonka E, Raz G (2009) Transgenerational epigenetic inheritance: prevalence, mechanisms, and implications for the study of heredity and evolution. Q Rev Biol 84:131–176

Janska A, Marsik P, Zelenkova S, Ovesna J (2010) Cold stress and acclimation—what is important for metabolic adjustment? Plant Biol 12:395–405

Jeon J, Chulo C, Lee MR, Binh NV, Kim J (2016) *CYTOKININ RESPONSE FACTOR2* (*CRF2*) and *CRF3* regulate lateral root development in response to cold stress in *Arabidopsis*. Plant Cell 28:1828–1843

Jones-Rhoades MW, Bartel DP, Bartel B (2006) MicroRNAS and their regulatory roles in plants. Annu Rev Plant Biol 57:19–53

Kalberer SR, Wisniewski M, Arora R (2006) Deacclimation and reacclimation of cold-hardy plants: current understandings and emerging concepts. Plant Sci 171:3–16

Kaniuga Z, Saczynska V, Miskiewicz E, Garstka M (1999) The fatty acid composition of phosphatidylglycerol and sulfoquinovosyldiacylglycerol of *Zea mays* genotypes differing in chilling susceptibility. J Plant Physiol 154:256–263

Kanaoka MM, Pillittere LJ, Fujii H, Yoshida Y, Bogenschutz NL, Takabayashi J, Zhu JK, Torii KU (2008) SCREAM/ICE1 and SCREAM2 specify three cell-state transitional steps leading to *Arabidopsis* stomatal differentiation. Plant Cell 20:1775–1785

Kaplan F, Kopka J, Haskell DW, Zhao W, Schiller KC, Gatzke N, Sung DY, Guy CL (2004) Exploring the temperature-stress metabolome of *Arabidopsis*. Plant Physiol 136:4159–4168

Kasamo K, Kagita F, Yamanishi H, Sasaki T (1992) Low temperature-induced changes in the thermotropic properties and fatty acid composition of the plasma membrane and tonoplast of cultured rice (*Oryza sativa* L) cells. Plant Cell Physiol 33:609–616

Kasamo K, Noushi I (1987) The role of phospholipids in plasma membrane ATPase activity in *Vigna radiata* L. (mung bean) roots and hypocotyls. Plant Physiol 83:323–328

Kasuga M, Miura S, Shinozaki K, Yamaguchi-Shinozaki K (2004) A combination of the *Arabidopsis DREB1A* gene and stress-inducible *rd29A* promoter improved drought- and low-temperature stress tolerance in tobacco by gene transfer. Plant Cell Physiol 45:346–350

Kasukabe Y, He L, Nada K, Misawa S, Ihara I, Tachibana S (2004) Overexpression of spermidine synthase enhances tolerance to multiple environmental stresses and up-regulates the expression of various stress-regulated genes in transgenic *Arabidopsis thaliana*. Plant Cell Physiol 45:712–722

Khraiwesh B, Zhu JK, Zhu J (2012) Role of miRNAs and siRNAs in biotic and abiotic stress responses of plants. Biochim Biophys Acta 1819:137–148

Kim JC, Lee SH, Cheong YH, Yoo CM, Lee SI, Chun HJ, Yun DJ, Hong JC, Lee SY, Lim CO, Cho MJ (2001) A novel cold- inducible zinc finger protein from soybean, SCOF-1, enhances cold tolerance in transgenic plants. Plant J 25:247–259

Kim TW, Wang ZY (2010) Brassinosteroid signal transduction from receptor kinases to transcription factors. Annu Rev Plant Biol 61:681–704

Knight H, Trewavas AJ, Knight MR (1996) Cold calcium signaling in *Arabidopsis* involves two cellular pools and a change in calcium signature after acclimation. Plant Cell 8:489–503

Kojima M, Suzuki H, Ohnishi M, Ito S (1998) Effects of growth temperature on lipids of adzuki bean cells. Phytochemistry 47:1483–1487

Kolesnichenko AV, Pobezhimova TP, Grabelnych OI, Voinikov VK (2002) Stress induced protein CSP 310: a third uncoupling system in plants. Planta 215:279–286

Kovács Z, Simon-Sarkadi L, Szûcs A, Kocsy G (2010) Differential effects of cold, osmotic stress and abscisic acid on polyamine accumulation in wheat. Amino Acids 38:623–631

Kovtun Y, Chiu WL, Tena G, Sheen J (2000) Functional analysis of oxidative stress-activated mitogen-activated protein kinase cascade in plants. Proc Natl Acad Sci U S A 97:2940–2945

Krupa Z, Huner NP, Williams JP, Maissan E, James DR (1987) Development at cold-hardening temperatures. The structure and composition of purified rye light harvesting complex II. Plant Physiol 84:19–24

Kumar SV, Wigge PA (2010) H2A.Z-containing nucleosomes mediate the thermosensory response in *Arabidopsis*. Cell 140:136–147

Lee HG, Lee K, Seo PJ (2015) The *Arabidopsis* MYB96 transcription factor plays a role in seed dormancy. Plant Mol Biol 87:371–381

Lee SH, Singh AP, Chung GC, Kim YS, Kong IB (2002) Chilling root temperature causes rapid ultrastructural changes in cortical cells of cucumber (*Cucumis sativus* L.) root tips. J Exp Bot 53:2225–2237

Lee SJ, Kang JY, Park HJ, Kim MD, Bae MS, Choi HI (2010) DREB2C interacts with ABF2, a bZIP protein regulating abscisic acid-responsive gene expression, and its overexpression affects abscisic acid sensitivity. Plant Physiol 153:716–727

Li M, Hada A, Sen P, Olufemi L, Hall MA, Smith BY, Forth S, McKnight JN, Patel A, Bowman GD, Bartholomew B, Wang MD (2015) Dynamic regulation of transcription factors by nucleosome remodeling. Elife. https://doi.org/10.7554/eLife.06249

Lin MK, Belanger H, Lee YJ, Varkonyi-Gasic E, Takoa K, Miura E, Xoconostle-Cazares B, Gendler K, Jorgensen RA, Phinney B, Lough TJ, Lucas WJ (2007) *FLOWERING LOCUS T* protein may act as the long-distance florigenic signal in the cucurbits. Plant Cell 19:488–1506

Liu F, Jensen CR, Andersen MN (2004) Drought stress effect on carbohydrate concentration in soybean leaves and pods during early reproductive development: its implication in altering pod set. Field Crops Res 86:1–13

Liu X, Secombe J (2015) The histone demethylase KDM5 activates gene expression by recognizing chromatin context through its PHD reader motif. Plant Cell Rep 13:2219–2231

Livingston DP, Hincha DK, Heyer AG (2009) Fructan and its relationship to abiotic stress tolerance in plants. Cell Mol Life Sci 66:2007–2023

Lukatkin AS, Brazaityte A, Bobinas C, Duchovskis P (2012) Chilling injury in chilling-sensitive plants: a review. Zemdirbyste-Agriculture 99:111–124

Mayer BF, Ali-Benali MA, Demone J, Bertrand A, Charron JB (2015) Cold acclimation induces distinctive changes in the chromatin state and transcript levels of *COR* genes in *Cannabis sativa* varieties with contrasting cold acclimation capacities. Physiol Plant 155:281–295

Miura K, Furumoto T (2013) Cold signaling and cold response in plants. Int J Mol Sci 14:5312–5337

Miura K, Jin JB, Lee J, Yoo CY, Stirm V, Miura T, Ashworth EN, Bressan RA, Yun DJ, Hasegawa PM (2007) SIZ1-mediated sumoylation of ICE1 controls CBF3/DREB1A expression and freezing tolerance in *Arabidopsis*. Plant Cell 19:1403–1414

Miura K, Lee J, Gong Q, Ma S, Jin JB, Yoo CY, Miura T, Sato A, Bohnert HJ, Hasegawa PM (2011) SIZ1 regulation of phosphate starvation-induced root architecture remodeling involves the control of auxin accumulation. Plant Physiol 155:1000–1012

Murata N, Mohanty PS, Hayashi H, Papageorgiou GC (1992) Glycinebetaine stabilizes the association of extrinsic proteins with the photosynthetic oxygen-evolving complex. FEBS Lett 296:187–189

Murphy C, Wilson JM (1981) Ultrastructural features of chilling-injury in *Episcia reptans*. Plant Cell Environ 4:261–265

Nakashima K, Ito Y, Yamaguchi-Shinozaki K (2009) Transcriptional regulatory networks in response to abiotic stresses in *Arabidopsis* and grasses. Plant Physiol 149:88–95

Nuener G, Xu B, Hacker J (2010) Velocity and pattern of ice propagation and deep supercooling in woody stems of *Castanea sativa, Morus nigra* and *Quercus robur* by IDTA. Tree Physiol 30:1037–1045

Oliveira G, Penuelas J (2005) Effects of winter cold stress on photosynthesis and photochemical efficiency of PSII of Mediterranean *Cistus albidus* L. and *Quercus ilex* L. Plant Ecol 175:179–191

Örvar BL, Sangwan V, Omann F, Dhindsa RS (2000) Early steps in cold sensing by plant cells: the role of actin cytoskeleton and membrane fluidity. Plant J 23:785–794

Paredes M, Quiles MJ (2015) The effects of cold stress on photosynthesis in *Hibiscus* plants. PLoS One 10:e0137472

Parkin KL, Marangoni A, Jackman RL, Yada RY, Stanley DW (1989) Chilling injury: a review of possible mechanism. J Food Biochem 13:1745–4514

Pavangadkar K, Thomashow MF, Triezenberg SJ (2010) Histone dynamics and roles of histone acetyltransferases during cold-induced gene regulation in *Arabidopsis*. Plant Mol Biol 74:183–200

Pearce RS (2001) Plant freezing and damage. Ann Bot 87:417–424

Pramanik MHR, Imai R (2005) Functional identification of a trehalose 6-phosphate phosphatase gene that is involved in transient induction of trehalose biosynthesis during chilling stress in rice. Plant Mol Biol 58:751–762

Ratcliffe OJ, Kumimoto RW, Wong BJ, Riechmann JL (2003) Analysis of the *Arabidopsis MADS AFFECTING FLOWERING* gene family: *MAF2* prevents vernalization by short periods of cold. Plant Cell 15:1159–1169

Ribas-Carbo M, Aroca R, Gonzàlez-Meler MA, Irigoyen JJ, Sánchez-Díaz M (2000) The electron partitioning between the cytochrome and alternative respiratory pathways during chilling recovery in two cultivars of maize differing in chilling sensitivity. Plant Physiol 122:199–204

Rihan HZ, Al-Issawi M, Fuller MP (2017) Advances in physiological and molecular aspects of cold tolerance. J Plant Interact 12:143–157

Rinne PL, Kaikuranta PM, van der Schoot C (2001) The shoot apical meristem restores its symplasmic organization during chilling-induced release from dormancy. Plant J 26:249–264

Rohde A, Bhalerao RP (2007) Plant dormancy in the perennial context. Trends Plant Sci 12:217–223

Rolland F, Baena-Gonzalez E, Sheen J (2006) Sugar sensing and signaling in plants: conserved and novel mechanisms. Annu Rev Plant Biol 57:675–709

Routaboul JM, Fischer S, Browse J (2000) Trienoic fatty acids are required for photosynthesis at low temperatures. Plant Physiol 124:1697–1705

Roy D, Paul A, Roy A, Ghosh R, Ganguly P, Chaudhuri S (2014) Differential acetylation of histone H3 at the regulatory region of OsDREB1b promoter facilitates chromatin remodelling and transcription activation during cold stress. PLoS One 9:e100343

Sakai A, Larcher W (1987) Frost survival of plants: responses and adaptations to freezing stress. Springer-Verlag, Berlin, pp 1–259

Sanghera GS, Wani SH, Hussain W, Singh NB (2011) Engineering cold stress tolerance in crop plants. Curr Genomics 12:30–43

Sangwan V, Foulds I, Singh J, Dhindsa RS (2001) Cold-activation of *Brassica napus* BN115 promoter is mediated by structural changes in the membrane and cytoskeleton, and requires Ca^{2+} influx. Plant J 27:1–12

Sarhan F, Ouellet F, VazquezTello A (1997) The wheat WCS120 gene family. A useful model to understand the molecular genetics of freezing tolerance in cereals. Physiol Plant 101:439–445

Sasse JM (2003) Physiological actions of brassinosteroids: an update. J Plant Growth Regul 22:276–288

Schmitz RJ, Sung S, Amasino RM (2008) Histone arginine methylation is required for vernalization-induced epigenetic silencing of FLC in winter-annual *Arabidopsis thaliana*. Proc Natl Acad Sci U S A 105:411–416

Shan X, Wang X, Yang G, Wu Y, Su S, Li S, Liu H, Yuan Y (2013) Analysis of the DNA methylation of maize (*Zea mays* L.) in response to cold stress based on methylation-sensitive amplified polymorphisms. J Plant Biol 56:32–38

Sheen J (1998) Mutational analysis of protein phosphatase 2C involved in abscisic acid signal transduction in higher plants. Proc Natl Acad Sci U S A 95:975–980
Shibasaki K, Uemura M, Tsurumi S, Rahman A (2009) Auxin response in *Arabidopsis* under cold stress: underlying molecular mechanisms. Plant Cell 21:3823–3838
Shindo C, Aranzana MJ, Lister C, Baxter C, Nicholls C, Nordborg M, Dean C (2005) Role of FRIGIDA and FLOWERING LOCUS C in determining variation in flowering time of *Arabidopsis*. Plant Physiol 138:1163–1173
Spano C, Bottega S, Castiglione MR, Pedranzani H (2017) Antioxidant response to cold stress in two oil plants of the genus *Jatropha*. Plant Soil Environ 63:271. https://doi.org/10.17221/182/2017-PSE
Stockinger EJ, Gilmour SJ, Thomashow MF (1997) *Arabidopsis thaliana* CBF1 encodes an AP2 domain-containing transcriptional activator that binds to the C-repeat/DRE, a cis-acting DNA regulatory element that stimulates transcription in response to low temperature and water deficit. Proc Natl Acad Sci U S A 94:1035–1040
Strange A, Li P, Lister C, Anderson J, Warthmann N, Shindo C, Irwin J, Nordborg M, Dean C (2011) Major-effect alleles at relatively few loci underlie distinct vernalization and flowering variation in *Arabidopsis* accessions. PLoS One 6:e1994
Strauss AJ, Krüger GHJ, Strasser RJ, Heerden PDR (2006) Ranking of dark chilling tolerance in soybean genotypes probed by the chlorophyll a fluorescence transient O-J-I-P. Environ Exp Bot 56:147–157
Sugie A, Naydenov N, Mizuno N, Nakamura C, Takumi S (2006) Overexpression of wheat alternative oxidase gene Waox1a alters respiration capacity and response to reactive oxygen species under low temperature in transgenic *Arabidopsis*. Genes Genet Syst 81:349–354
Sung S, Schmitz RJ, Amasino RM (2006) A PHD finger protein involved in both the vernalization and photoperiod pathways in *Arabidopsis*. Genes Dev 20:3244–3248
Svensson AJ, Johansson FI Moller IM, Rasmusson AG (2002) Cold stress decreases the capacity for respiratory NADH oxidation in potato leaves. FEBS Lett 517:79–82
Sweetlove LJ, Lytovchenko A, Morgan M, Nunes-Nesi A, Taylor NL, Baxter CJ, Eickmeier I, Fernie AR (2006) Mitochondrial uncoupling protein is required for efficient photosynthesis. Proc Natl Acad Sci U S A 103:19587–19592
Talts P, Parnik T, Gardeström P, Keerberg O (2004) Respiratory acclimation in *Arabidopsis thaliana* leaves at low temperature. J Plant Physiol 161:573–579
Tang Z, Zhang L, Xu C, Yuan S, Zhang F, Zheng Y, Zhao C (2012) Uncovering small RNA-mediated responses to cold stress in a wheat thermosensitive genic male-sterile line by deep sequencing. Plant Phyisiol 159:721–738
Taylor NL, Heazlewood JL, Millar AH (2011) The *Arabidopsis thaliana* 2-D gel mitochondrial proteome: refining the value of reference maps for assessing protein abundance, contaminants and post-translational modifications. Proteomics 11:1720–1733
Thiebaut F, Grativol C, Carnavale-Bottino M, Rojas CA, Tanurdzic M, Farinelli L, Martienssen RA, Hemerly AS, Ferreira PCG (2012) Computational identification and analysis of novel sugarcane microRNAs. BMC Genomics 13:290
Thomashow MF (1999) Plant cold acclimation: freezing tolerance genes and regulatory mechanisms. Annu Rev Plant Physiol Plant Mol Biol 50:571–599
Vella NGF, Joss TV, Roberts TH (2012) Chilling-induced ultrastructural changes to mesophyll cells of *Arabidopsis* grown under short days are almost completely reversible by plant re-warming. Protoplasma 249:1137–1149
Vijayakumar H, Thamilasaran SK, Shanmugam A, Natarajan S, Jung H, Park J, Kim H, Chung M, Nou I (2016) Glutathione transferases superfamily: cold-inducible expression of distinct *GST* genes in *Brassica oleracea*. Int J Mol Sci 17:1211
Whitaker BD (1993) Lipid changes in bell pepper fruit during chilling and rewarming. Hortscience 28:519
Vlachonasios KE, Thomashow MF, Triezenberg SJ (2003) Disruption mutations of *ADA2b* and *GCN5* transcriptional adaptor genes dramatically affect *Arabidopsis* growth, development, and gene expression. Plant Cell 15:626–638

Wan B, Lin Y, Mou T (2007) Expression of rice Ca(2+)-dependent protein kinases (CDPKs) genes under different environmental stresses. FEBS Lett 581:1179–1189

Wani SH, Sandhu JS, Gosal SS (2008) Genetic engineering of crop plants for abiotic stress tolerance. In: Malik CP, Kaur B, Wadhwani C (eds) Advanced topics in plant biotechnology and plant biology. MD Publications, New Delhi, pp 149–183

Wang Q, Liu N, Yang X, Tu L, Zhang X (2016) Small RNA-mediated responses to low- and high-temperature stresses in cotton. Sci Rep 6:35558. https://doi.org/10.1038/srep35558

Wisniewski M, Fuller M, Palta J, Arora R (2008) Ice nucleation, propagation and deep supercooling in woody plants. J Crop Improvement 10(1–2):2004

Wi SJ, Kim WT, Park KY (2006) Overexpression of carnation *S*-adenosylmethionine decarboxylase gene generates a broad spectrum tolerance to abiotic stresses in transgenic tobacco plants. Plant Cell Rep 25:1111–1121

Xiang Y, Huang Y, Xiong L (2007) Characterization of stress-responsive CIPK genes in rice for stress tolerance improvement. Plant Physiol 144:1416–1428

Xin Z, Browse J (1998) Eskimo1 mutants of *Arabidopsis* are constitutively freezing-tolerant. Proc Natl Acad Sci U S A 95:7799–7804

Xiong LZ, Yang YN (2003) Disease resistance and abiotic stress tolerance in rice are inversely modulated by an abscisic acid-inducible mitogen-activated protein kinase. Plant Cell 15:745–759

Yang T, Chaudhuri S, Yang L, Du L, Poovaiah BW (2012) A calcium/calmodulin-regulated member of the receptor-like kinase family confers cold tolerance in plants. J Biol Chem 285:7119–7126

Yin PH, Wu CC, Lin JC, Chi CW, Wei YH, Lee HC (2010) Somatic mutations of mitochondrial genome in hepatocellular carcinoma. Mitochondrion 10:174–182

Zhang J, Xue B, Gai M, Song S, Jia N, Sun H (2017) Small RNA and transcriptome sequencing reveal a potential miRNA-mediated interaction network that functions during somatic embryogenesis in *Lilium pumilum* DC. Fisch. Front Plant Sci. https://doi.org/10.3389/fpls.2017.00566

Zhao M, Liu W, Xia X, Wang T, Zhang WH (2014) Cold acclimation-induced freezing tolerance of *Medicago truncatula* seedlings is negatively regulated by ethylene. Physiol Plant 152:115–129

Zhou X, Chen S, Wu H, Xu H (2017) Effects of cold stress on the photosynthesis and antioxidant system of *Rhododendron chrysanthemum* Pall. www.preprints.org. doi:https://doi.org/10.20944/preprints201703.0131.v1

Zhu J, Shi H, Lee BH, Damsz B, Cheng S, Stirm V, Zhu JK, Hasegawa PM, Bressan RA (2004) An *Arabidopsis* homeodomain transcription factor gene, HOS9, mediates cold tolerance through a CBF-independent pathway. Proc Natl Acad Sci U S A 101:9873–9878

Chapter 2
Molecular Genetic Approaches for the Identification of Candidate Cold Stress Tolerance Genes

Muhammad Qudrat Ullah Farooqi, Zahra Zahra, and Ju Kyong Lee

2.1 Introduction

Abiotic stresses cause severe losses in worldwide crop production (Mittler 2006). Cold is one of most important stresses responsible for the reduction in plant yield especially in subtropical and temperate grain crops. Low-temperature stress causes considerable loss of agricultural crop yield, particularly in sensitive crops like maize, rice, and chickpea. Low temperatures delay flowering—the onset of reproduction. The reproductive phase particularly meiosis is more sensitive than the vegetative, while low temperature causes more damage to male reproductive organs than to female organs. The entire processes from gamete formation to fertilization and maturation of seeds are susceptible to cold stress. Most of the temperate region plants require freezing tolerance in the process called as cold acclimation. The basic important step to overcome the chilling stress in plant is to develop good phenotyping frameworks. It is essential for farmers and breeders to apply suitable phenotyping methodologies to control initial damage due to stress (Salekdeh et al. 2009). Liu et al. (1998) and Karaba et al. (2007) revealed that the environment is a basic factor to increase the chance of stress mechanism in crop plant.

M. Q. U. Farooqi (✉)
Department of Soil Science, School of Agriculture and Environment, Faculty of Science, The University of Western Australia, Perth, WA, Australia
e-mail: muhammadqudratullah.farooqi@uwa.edu.au

Z. Zahra
Institute of Environmental Sciences and Engineering, School of Civil and Environmental Engineering, National University of Sciences and Technology, Islamabad, Pakistan

J. K. Lee
Department of Applied Plant Sciences, College of Agriculture and Life Sciences, Kangwon National University, Chuncheon, Republic of Korea

S. H. Wani, V. Herath (eds.), *Cold Tolerance in Plants*,
https://doi.org/10.1007/978-3-030-01415-5_2

The environmental variation can be minimized by selecting experimental sites under controlled conditions such as pots, soil-filled pipes for water drainage, and hydroponics. However, controlled conditions tend to be difficult to attain higher yield (Hu et al. 2006). For this, identification of cold stress phenotyping sites is very important. Molecular breeding and geographical information system (GIS) are main tools to improve the groundwork for phenotyping in plants (Okono et al. 2014). The physiological adjustment and adaptation mechanism help crop plant to tolerate or avoid against stress. Plants have adapted various mechanisms to ensure survival under stress (Mickelbart et al. 2015). The stress tolerance mechanisms are caused by cellular and biochemical modifications by which the plant can survive against harsh conditions. In contrast, stress avoidance mechanism results due to changes in physiological and morphological characters in the entire plant; due to this, the plant is unable to survive and resists growing (Subbaiah and Sachs 2003). The genetic diversity is an essential investigation for environmental adaptation and acclimation of model plant species.

Several abiotic stresses disturb plant metabolic activities such as overproduction of reactive oxidation species (ROS) (Mittler 2002). Reactive oxidation species causes severe damage in plant metabolism, for example, the amount of carbohydrates, lipids, proteins, and even in DNA of crop plant during stress (Gill and Tuteja 2010). In defensive mechanism, the plant induces proline which is an important protein to reduce the impact of cold stress (Hare and Cress 1997). Cold stress causes freezing injuries, cellular damage (necrosis), and disturbing water potential gradient in crop plant (Farooqi et al. 2016; Guy 1990). The proportion of protein, starch, and oil contents significantly reduces during grain-filling stage. The glucokinase, sucrose synthase, and ADP glucose pyrophosphorylase are the important enzymes at developmental stages. The ADP glucose pyrophosphorylase activity is severely affected under stress conditions (Wilhelm et al. 1998).

The identification of quantitative trait loci (QTL) is important for the detection of genomic region associated with stress tolerance traits. Quantitative trait loci are important to configure physiological mechanism and to control heritable variability due to stress. QTL can be detected in specific environmental conditions or strongly expressed with the level of environmental factor (Collins et al. 2008). Crop performance and yield improvement can be better analyzed with the findings of stress-linked specific loci under environmentally constrained conditions (Rodriguez et al. 2014). Therefore, the genetic dissection of stress-linked quantitative loci is necessary to improve the stability and sustainability of crop yield under environmentally severe conditions (Welcker et al. 2007).

Several comprehensive studies have been published recently about cold stress response mainly largely in associated genes, signal transduction system, and genetic engineering of abiotic stress tolerance in plants (Seki et al. 2003). However, the phenomenon for cold stress adaptation is yet poorly understood. In the chapter, we have selected recent development of cold stress tolerance that is presented in our understanding with special emphasis on the interacting factors in molecular biology for plants' tolerance to cold stress.

2.2 Cold Stress Tolerance Mechanism

Cold temperature stress (CTS) causes dysfunction on several physiological, morphological, and biochemical processes such as photosynthesis, water and mineral intake, respiration, and fresh and dry biomass. A schematic diagram, the cold stress response in crop plant is demonstrated (Fig. 2.1). CTS stimulates an imbalance between uptake and transport of water contents and the generation of reactive oxygen species (ROS) (Mittler 2002). CTS adversely affects the growth of young seedlings and probably predisposes the plants to invasion by soil fungi capable of causing seed rot and seedling blight. CTS stress also affects the germination rate and if accompanied by precipitation caused by snow and freezing rain causes irreparable damages to seedlings. In addition, restricted growth causes sugar-induced anthocyanin accumulations in leaves which can cause purple coloration of leaves in many plants (Rao et al. 2017).

Chilling temperature reduces the photosynthetic capacity that is one of the major reasons for the lower photosynthetic capacity in the perturbation of chloroplast development. Lower temperature developments lower the efficiency to develop functional photosynthetic apparatus. Ying et al. (2011) reported that in cold nights, cereals differ characteristically during their grain-filling and flowering stages which in turn affects the grain-filling mechanism and its effectiveness. However, genetic variation does exist for the severity of the response to low temperatures for several

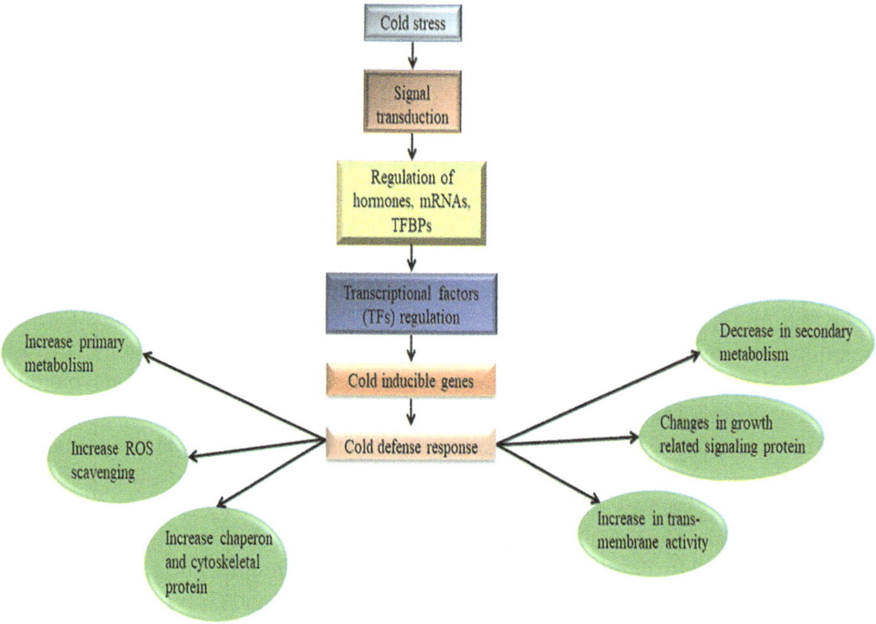

Fig. 2.1 A schematic diagram of cold stress response in plants

parameters, such as rate of development, leaf CER, the rate of dry matter accumulation, and quantum efficiency of PSII (Ying et al. 2011).

When plants are exposed to a sudden cold stress, the seedlings exhibit symptoms of drought stress due to an imbalance between transpiration and water uptake. In particular, ineffective stomatal control is observed shortly after the onset of cold stress, which is the result of a decrease in the hydraulic conductance of the roots due to the greater viscosity of water at low temperature and to intrinsic characteristics of the root (Melkonian et al. 2004). The physiological activity of roots of chill-sensitive genotypes was marked by a stronger inhibition of water uptake and root respiration than chill-tolerant genotypes when exposed to chilling temperature (Sowinski and Maleszewski 1989). The chill-induced change in the water status induced the generation of abscisic acid (ABA), especially in chill-tolerant maize genotypes as was demonstrated under controlled conditions as well as in the field following a cold spell.

2.3 Plant Metabolic Response to Cold Tolerance

Plant shows various metabolic responses under chilling stress. In severe conditions, membrane permeability, photosynthetic rate, and osmotic potential have been deeply affected. The clarification of metabolic activities in response to low temperature is essential for the optimization of maize cultivation and breeding to obtain high yield. Various metabolites are involved to control plant growth under different kinds of abiotic stresses.

In response to environmental stress, the reactive oxygen species (ROS) level is dramatically increased which is harmful for plant survival. Plant produces defensive mechanism to reduce the effect of low temperature. The accumulation and production of compatible solutes had been observed in many studies under stress conditions. These compatible solutes are osmoprotectants which include amino acids (proline, serine), amines (polyamines), and γ-amino-N-butyric acid (GABA). Moreover, carbohydrates are major pools for antioxidant (glutathione and ascorbate) accumulation due to osmotic stress. Raffinose is another important carbohydrate that protects plant from oxidative damage. The osmoprotectants have various functions such as protection and stability of cellular structures and proteins, diffuse reactive oxygen species (ROS) to redox plant metabolism, and osmotic balance (Rathiasababapathi 2000; Mittler 2002).

Temperature stress especially freezing environment leads to serious damage in plant cell membrane due to ice formation. In adverse, the translation of heat shock proteins, kinases and phosphatases increased during heat stress. High exposure of light also affects plant metabolic activities. Intense light absorption and its exposure to plant leaves lead to far-red wavelength enrichment, which imbalances photosystems I and II during transport chain and Calvin-Benson cycle (Schluter et al. 2013).

2.4 Effect on the Reproductive Development Due to Cold Stress

In plant life cycle, vegetative and reproductive phases are deeply affected by cold stress (Nishiyama 1995). Low-temperature stress during the reproductive phase induces flower abscission, pollen sterility, pollen tube abortion, ovule distortion, and reduced fruit set, which ultimately reduced yield. The reproductive phase begins with the transformation of the meristem into inflorescence and flower and, in annuals, ends upon seed reaching maturity. All of the sequential stages of the reproductive phase respond differently to cold stress (Staggenborg and Vanderlip 1996; Verheul et al. 1996); but collectively all responses are negative and cause reduction in net yield. Thus, cold stress during the reproductive phase has important socioeconomic consequences because the products of the reproductive phase are key components of economic yield and are the principal source of food around worldwide.

Cold stress causes structural and functional abnormalities in reproductive organs, leading to failure of fertilization or premature abortion of seed or fruit during the reproductive phase. The literature on chilling stress during the reproductive phase dominated by rice is very sensitive. Farooq et al. (2009) have reviewed recently the agronomic and physiological responses of chilling stress in maize. Chilling-sensitive plants exposed to low temperature often show signs of water stress due to lowered root hydraulic conductance leading to lowering in leaf water and turgor potential, followed by a reduction or completely cessation of growth (Aroca et al. 2001). Initially this phenomenon is reversible but ultimately becomes irreversible and can result in cell death. The degree of injury associated with chilling temperatures varies with plant species, stage of crop development, rate of change of temperature, duration of exposure, irradiance, and mineral nutrition (Ercoli et al. 2004).

In plants, the cold stress perception is connected to changes in plasma membrane fluidity, triggering a raft of downstream effects including the activation of signaling pathways involving ABA and other important hormones. It appears to be a common plant response to cold stress across the entire reproductive cycle by changing carbohydrate metabolism under ABA-mediated regulation. ABA accumulates in aborted flowers in response to cold, suppresses the transcription of anther-specific invertase genes leading to disturbed carbohydrate metabolism and hypertrophy, and may play a role in modifying pollen tube growth by diverting carbohydrate supply in the style (Feng et al. 2009).

2.5 Plant Genetics of Cold Stress Tolerance and Avoidance

Plant shows tolerance or avoidance to stress mechanism through adaptation that has evolved by natural selection. Abiotic stresses especially cold, salinity, and drought have major impact on crop production (Wang et al. 2003). Until recently, the mechanism by which plant withstands under cold stress is poorly understood. Natural variation in crop plant clears the mechanism of stress tolerance. For this, genetic analyses of stress-linked traits are essential, and many candidate genes are available for crop improvement. It is necessary to have accurate and well-defined genetic material to identify the evidences behind stress tolerance. Another important thing is to know about genetic and environmental effect (GXE) because the environment is a basic factor to create genetic variation. The genetic studies of cold stress tolerance have significant challenge to increase yield and speed at which phenotyping assays can be made (Richards et al. 2010; Masuka et al. 2012).

Genetic variation is very important for crop improvement. The crop yield improvement is closely associated with improved stress tolerance. The combination of different abiotic stresses could be linked genetically to suppress plant growth or yield. Crossing of diversified inbred lines with specific combination can be effective for the selection of desirable stress-linked traits. General combining ability (GCA) and specific combining ability (SCA) describe the breeding values of parental lines to evolve new hybrids. The ratio between general and specific combining ability is useful to estimate genetic variability. To understand cold stress response in any plant, it is very important to know the genetic studies of roots. It is due to roots immediate response against any unfavorable condition (Khodarahmpour 2011).

The genetic approach of cold stress tolerance is a complex phenomenon under cold or drought condition. Although, reverse genetics based on candidate gene can adversely affect stress tolerance mechanism in crop plant. However, a more powerful technique is required to identify the variation due naturally occurring cold stress tolerance in plant genotypes. The genetic loci used to determine stress specific traits by correlating the genetic mutation and the trait values in large mapping population (Meseka et al. 2011). Once the molecular bases of these traits have been identified, it is easy to implement genetic modification (GM) and marker-assisted breeding technologies for high yielding capacity under cold stress. In this way, some important technologies are implemented to get rid of conventional phenotyping method like nondestructive imaging (NDT). Nondestructive imaging has already been used to quantify traits related to cold, drought, heat, and salt tolerance in crop plant (Berger et al. 2010).

2.6 Genetic Engineering of Cold Stress Traits

Genetic engineering is a quicker way to insert useful genes responsive to abiotic stress tolerance than molecular breeding. Genetic loci that provide performance in stressful environment can exist within the germplasm of crop plant. These loci are often associated with separate duplication or regulation of plant gene functions to maintain plant homeostasis (Takeda and Matsuoka 2008). Several stress-induced genes are introduced by genetic engineering in maize plant to increase tolerance to cold, drought, and salt stresses. Biotechnology offers new strategies for the development of transgenic crop plants with improvement in their surviving mechanism to cold stress. Advancing in genetic engineering technologies like recombinant

DNA and efficient gene transfer protocol resulted in better transformation and development of excellent transgenic lines in various numbers of crop species (Wani et al. 2008).

The advancement in molecular biology tools is useful for the regulation of cold stress tolerance mechanism based on expression analysis of different stress-linked genes. A variety of cold stress-linked genes have been identified to reconstruct stress tolerance in plants. The data collected for the genetic engineering is based on high-throughput transcription profiling, molecular modeling, defensive mechanism against antioxidants, and the identification of specific protein network in high scale against cold stress and various climatic changes in plants (Jangra et al. 2017). It is very important to protect the plant cell from the damage of enzymatic antioxidants due to cold stress. For example, glycine betaine is a key enzyme to improve cold stress resistance. It has been identified that the enhanced level of glycine betaine is effective for the stress tolerance in many plants. Moreover, glycine betaine impacts on several endogenous gene expressions in transgenic plants (Kumar et al. 2017).

Basic progress has been made in model plant *Arabidopsis* and winter cereals to unwind the molecular basis of cold acclimation. Plants regulate their gene expression through posttranslational, transcriptional, and posttranscriptional mechanisms during cold stress. Recently, various components of CBF transcriptional pathways of cold acclimation have been identified. The genetic engineering of CBF pathway can improve cold resistance in many species due to transgenic analysis. In a diverse range of plant species, the ICE1-CBF transcription factor plays an important role to withstand against chilling stress. Plants employ with diversified posttranscriptional mechanism for the regulation of gene expression during cold stress. In rice and *Arabidopsis*, a wide range of microRNA has been identified which helps to regulate the stability of mRNA in response to cold acclimation. The process of epigenetic plays an important role for the DNA methylation and chromatin modification. Most of transcriptional cold mechanisms were studied during the vegetative stages of plants. However, cold-induced transcriptomes are significantly different among leaf root and reproductive tissues (Chinnusamy et al. 2011).

2.7 Regulation of Gene Expression

Plants respond with changes in their metabolic pathways and gene expression in gene pattern when exposed to freezing conditions. Most of the gene required cis-, trans- regulatory features altered expression levels during abiotic stress. Gene expression pattern and protein products can be altered due to stress. Many genome-wide association projects related to cold stress have been completed to identify stress-linked genes in plants. GeneChips or cDNA microarrays are important tools for transcriptome analysis to discover genes involved in stress tolerance (Zheng et al. 2010).Genes are probably expressed and functionally response in stress. Many efficient methods are widely used for the expression of gene such as suppression subtractive hybridization (SSH) (Zheng et al. 2004). Transposable elements can contribute to gene activation in cold stress response (Makarevitch et al. 2015). Recently, cDNA microarray is a powerful tool in profiling of gene expression (Ji et al. 2003). Complementary DNA and suppression subtractive hybridization were combined and used successfully to identify differential gene expression (Yang et al. 1999).

There are several expressed genes in various plants for cold acclimation (Listed in Table 2.1). In apple and *Arabidopsis*, MdYH5 regulate the expression of CBF-independent cold-regulated genes (An et al. 2017). Legrand et al. (2013) combined the expression of 159 cold-regulated genes involved in freezing response of pea. In cucumber, the overexpression of *CsWRKY46* gene regulates chilling resistance and ABA hormone treatment by binding *ABI5* promoter (Zhang et al. 2016). CBADH1 gene is induced by cold tolerance in crossflower plant which contains three motifs with specified N-terminal sequence (Liu et al. 2017). Cold acclimation in stiff brome is generated and characterized by the expression of *BdCBF* gene which regulates CBF regulon (Hao et al. 2017). The overexpression of *IbCBF3* gene enhances temperature stress resistance in sweet potato (Jin et al. 2017). The gene *SpGR* plays an important role to regulate glutathione reductase enzyme and for the scavenging of ROS species in needle grass under cold stress (Wang et al. 2017). Some related genes like alternative oxidase (AOX) are important to control oxidative damage and to maintain mitochondrial respiratory properties due to cold stress in plant, i.e., in chickpea (Moalem et al. 2017). The expression of *PsCor413im1* gene was induced by ABA and cold tolerance in mountain phlox and *Arabidopsis* (Zhou et al. 2017).

In higher plants, COR, KIN, LTI, and RD are cold-regulated and inducible genes which help to improve cell metabolism to adopt low temperature (Barrero-Sicilia et al. 2017). The *PstTPS1* gene is overexpressed in wheat against yellow rust and its hyphae expansion due to chilling environment (Zou et al. 2017). Gu et al. (2017) found that the expression of *RdreB1BI*gene in strawberry had enhanced cold resistance. The gene from nucleotidyl transferase protein (NTP) is expressed in different plant tissues to lower cold stress at developmental stage in rice plant (Yang et al. 2017). Some proteins are essential to induce C- repeat binding transcription factor (CBF) for lower cold acclimation like MPK3- and MPK6- in *Arabidopsis* (Li et al. 2017a, b, c). In barley plant, cold-induced gene CISP encodes small basic protein like CISP1 transcripts localized on root tips and primordium (Ying and Kidou 2017). Xu et al. (2017) identified teosinte branched1 (TCP) genes among 41 sequenced plants as most of these were involved in cold treatment.

The photoperiodic response gradually decreases due to cold acclimation like in the alfalfa plant; genes like SuSy, SPS, GaS, and GAPDH are responsible for carbohydrate metabolism under freezing temperature (Bertrand et al. 2017). In oil palm, EgDREB1 gene was identified as functional regulator in enhancing cold resistance and defensive mechanism against heat shock proteins (Azzeme et al. 2017). The overexpression of cold-responsive gene can be activated by applying phenolic acid like salicylic acid in grapes which caused the expression of VvCBF4 gene to minimize the effect of freezing temperature (Aazami and Mahna 2017). Su et al. (2017) found that the overexpression of GhDof1 could improve cold tolerance and seed oil

Table 2.1 Important candidate genes identified in response to cold stress in different plants

Gene product/ ID	Definition/function	Technique	Plant	Source
MdHY5	Regulate transcriptional and posttranscriptional level	Transient expression assays	Apple, *Arabidopsis*	An et al. (2017)
CsWRKY46	Cold and ABA treatment	Transient expression assays	Cucumber	Zhang et al. (2016)
CbADH1	Expressed in young leaves and epigenetic modification	ADH activity assay	Crossflower	Liu et al. (2017)
BdCBF	Characterized of RNAi mutants and regulate CBF regulon	Genome-wide transcriptome profiling	Stiff brome	Hao et al. (2017)
IbCBF3	Binds to CRT/DRE in the promoters of *COR* genes	Gene overexpression	Sweet potato	Jin et al. (2017)
SpGR	Scavenging of ROS species	Glutathione reductase content analyses	Needle grass	Wang et al. (2017)
PsCor413im1	Chloroplast membrane protein	Gene overexpression	Mountain phlox, *Arabidopsis*	Zhou et al. (2017)
SlICE, SlCBF, SlP5CS	Promote photosynthetic carbon fixation	Gene expression	Tomato	Ding et al. (2017)
LpCYP72A161	Demethylation on exon 1	Gene expression	Ryegrass	Dai et al. (2017)
COR, KIN, LTI, RD	Regulate functional proteins that intervene in cell metabolism	Transient expression assays	Saltwater cress, higher plants	Barrero-Sicilia et al. (2017)
PstTPS1	Pathogenicity by limiting hyphae expansion.	Gene expression	Wheat	Zou et al. (2017)
RdreB1BI	Bind the promoter of *FvPIP2 and AQP-related genes*	Synteny analysis	Strawberry	Gu et al. (2017)
NTPs	Conserved protein in different plant tissues	Gene expression	Rice	Yang et al. (2017)
MPK3, MPK6	Activate attenuates plant freezing tolerance and destabilize the ICE1 protein	Gene expression	*Arabidopsis*	Li et al. (2017a, b, c)
CISP	Bind RNA assay and maintain root primordium	RNA binding assay	Barley	Ying and Kidou (2017)
TCP, AtTCPs	Regulate developmental processes, i.e., branching and leaf growth	Transcriptome analysis	*Arabidopsis*, cereals, grasses	Xu et al. (2017)

(continued)

Table 2.1 (continued)

Gene product/ ID	Definition/function	Technique	Plant	Source
RING-HC, *ZmRHCP1*	Root development	Gene expression	Maize	Li et al. (2017a, b, c)
SlCOR413IM1	Maintain maximum photochemical efficiency	Transcriptome analysis	Tobacco	Ma et al. (2017)
bHLHs	Regulate many complex associations related with regular metabolic functions	Protein expression	Soybean	Filiz et al. (2017)
AOX3, *COX*, *SDH*	Control oxidative damage and maintain mitochondrial properties	Gene expression	Chickpea	Moalem et al. (2017)
SuSy, SPS, GaS, GAPDH	Carbohydrate metabolic response	Transcriptome analysis	Alfalfa	Bertrand et al. (2017)
EgDREB1	Initiation of signaling communication from root to shoot	Protein expression	Oil palm	Azzeme et al. (2017)
VvCBF4	Salicylic acid exogenous effect to increase cold tolerance	Gene expression	Grapes	Aazami and Mahna (2017)
GhDof1	Expressed in leaves, roots, and stems to improve cold stress	Transient expression assays	Cotton	Su et al. (2017)
AI5L2/ABF3	The improvement of exogenous ABA	Transcriptome analysis	Zucchini	Carvajal et al. (2017)
HXKs, FRKs	Hexokinases and fructokinases for carbohydrate metabolism	Gene expression	Tea	Li et al. (2017a, b, c)
MaERF10	Repression of biosynthetic gene	Transient expression assays	Banana	Qi et al. (2017)
PpFEH and *PpCAP*	Induced during cold and fructan accumulation	Cloning and sequence analysis	Smooth meadow-grass	Rao et al. (2017)
MtJMJC5	Response to circadian clock-induced alternative splicing	RNA binding assay	Barrel clover	Shen et al. (2017)
CmCBF1 and CmCBF3	Responsive to activate antioxidant enzyme on chilling injury	RNA binding assay	Hami melon	Zhang et al. (2017)
CBF/DREB1	Upregulation of cold tolerance in artificial seeds	Gene expression	Cauliflower	Rihan et al., 2017
*BoCRG*s	Regulate cold tolerance	Microarray expression analysis	Brassica	Ahmed et al. (2015)

content in cotton. In zucchini, AI5L2/ABF3 genes were responsible to improve cold tolerance during postharvest storage (Carvajal et al. 2017).

Li et al. (2017a, b, c) found hexose kinase genes (HXKs and FRKs) to regulate carbohydrate metabolism during low temperature in tea plant. The application of methyl jasmonate in banana was repressed due biosynthetic genes like MaERF10 in banana to mediate cold tolerance (Qi et al. 2017). Rao et al. (2017) revealed that the candidate genes (*PpFEH* and *PpCAP*) were involved in mediating freezing tolerance by regulating the cell osmotic potential through fructan degradation in meadow grasses. Shen et al. (2017) investigated that in barrel clover the *MtJMJC5* gene was involved in response to the circadian clock induced by alternate splicing during cold response. In tomato plant, genes like *SlICE*, *SlCBF*, and *SlP5CS* are essential to activate melatonin to improve antioxidant potential and carbon fixation rate during cold stress (Ding et al. 2017). The *LpCYP72A161* gene in ryegrass was expressed due to epigenetic modification and upregulated under cold stress which resulted in demethylation on exon 1 (Dai et al. 2017).

The chilling injury in plant can be reduced by nitric acid treatment, as it can express genes which are responsible in controlling chilling stress. Like in CmCBF1 and CmCBF3, genes in Hami melon were expressed to activate antioxidant enzyme on chilling injury (Zhang et al. 2017). *CBF* (core-binding factor) gene is not only important to upregulate cold stimulation in *Arabidopsis* but also in cauliflower mature plant (Rihan et al. 2017). Ahmed et al. (2015) found that the expression of *BoCRG*s gene was highly upregulated in the *Brassica* plant. Some genes are expressed in roots as roots are essential parts of plant to support against harsh environmental conditions. Like RING-HC gene, *ZmRHCP1* genes were involved in brace root development in maize plant in chilling conditions (Li et al. 2017a, b, c). *SlCOR413IM1* is another important gene that was expressed in tobacco plant to maintain photochemical efficacy (Ma et al. 2017). In soybean plant, the network of basic helix-loop-helix (*bHLH*) transcription factor genes in soybean was coexpressed to regulate various metabolic functions under cold stress (Filiz et al. 2017).

2.8 Conclusion

Plant biologists aimed to overcome environmental stresses using physiological, genetic engineering, and intensive molecular breeding technique at forefront. Various factors should be kept in mind while stress tolerance-related mechanism is under study. The exposure of stress to crop plant could be short or long span during molecular and physiological responses. However, most of the researchers use short-term stress treatment to get valuable results. Plant tolerance from stress has its molecular basis with the prevalent processes such as stress cycle and recovery from stress occurring under natural condition during various seasons. Thus, the comparative studies of expression and function of gene families in extreme condition will assist to minimize cold stress in plants.

References

Aazami MA, Mahna N (2017) Salicylic acid affects the expression of *VvCBF4* gene in grapes subjected to low temperature. J Genet Eng Biotechnol 15(1):257–261

Ahmed NU, Jung HJ, Park JI, Cho YG, Hur Y, Nou S (2015) Identification and expression analysis of cold and freezing stress responsive genes of *Brassica oleracea*. Gene 554:215–223

An JP, Yao JF, Wang XN, You CX, Wang XF, Hao YJ (2017) MdHY5 positively regulates cold tolerance via CBF-dependent and CBF-independent pathways in apple. J Plant Physiol 218:275–281

Aroca R, Tognoni F, Irigoyen JJ, Sánchez-Diaz M, Pardossi A (2001) Difference in root low temperature response of two maize genotypes differing in chilling sensitivity. Plant Physiol Biochem 39:1067–1075

Azzeme AM, Abdullah SNA, Aziz MA, Wahab PEM (2017) Oil palm drought inducible *DREB1* induced expression of DRE/CRT- and non-DRE/CRT-containing genes in lowland transgenic tomato under cold and PEG treatments. Plant Physiol Biochem 112:129–151

Barrero-Sicilia C, Silvestre S, Haslam RP, Michaelson LV (2017) Lipid remodelling: unravelling the response to cold stress in *Arabidopsis* and its extremophile relative *Eutremasalsugineum*. Plant Sci 263:194–200

Berger B, Parent B, Tester M (2010) High thoroughput shoot imaging to study drought responses. J Exp Bot 61:3519–3528

Bertrand A, Bipfubusa M, Claessens A, Rocher S, Castonguay Y (2017) Effect of photoperiod prior to cold acclimation on freezing tolerance and carbohydrate metabolism in alfalfa (*Medicago sativa* L.). Plant Sci 264:122–128

Carvajal F, Palma F, Munoz RJ, Jamilena M, Pulido A, Garrido D (2017) Unravelling the role of abscisic acid in chilling tolerance of zucchini during postharvest cold storage. Postharvest Biol Technol 133:26–35

Chinnusamy V, Zhu JK, Sunkar R (2011) Gene regulation during cold stress acclimation in plants. Methods Mol Biol 639:39–55

Collins NC, Tardieu F, Tuberosa R (2008) Quantitative trait loci and crop performance under abiotic stress: where do we stand? Plant Physiol 147:469–486

Dai Y, Mao P, Tao X, Wang Y, Wei CM, Ma XR (2017) Flexible shift on gene body methylation and transcription of *LpCYP72A161* exposed to temperature stress in perennial ryegrass. Environ Exp Bot 143:29–37

Ding F, Liu B, Zhang S (2017) Exogenous melatonin ameliorates cold-induced damage in tomato plants. Sci Hortic 219:264–271

Ercoli L, Mariotti M, Masoni A, Arduini I (2004) Growth responses of sorghum plants to chilling temperature and duration of exposure. Eur J Agron 21:93–103

Farooq M, Aziz T, Wahid A, Lee DJ, Siddique KHM (2009) Chilling tolerance in maize: agronomic and physiological approaches. Crop Pasture Sci 60:501–514

Farooqi MQU, Sa KJ, Hong TK, Lee JK (2016) Bulk segregant analysis (BSA) for improving cold stress resistance in maize using SSR markers. Genet Mol Res 15. https://doi.org/10.4238/gmr15049326

Feng DR, Liu B, Li WY, He YM, Qi KB, Wang HB, Wang JF (2009) Overexpression of a cold-induced plasma membrane protein gene (MpRCI) from plantain enhances low temperature-resistance in transgenic tobacco. Environ Exp Bot 65:395–402

Filiz E, Vatansever R, Ozyigit II (2017) Dissecting a co-expression network of basic helix-loop-helix (*bHLH*) genes from phosphate (pi)-starved soybean (*Glycine max*). Plant Gene 9:19–25

Gill SS, Tuteja N (2010) Reactive oxygen species and antioxidant machinery in abiotic stress tolerance in crop plants. Plant Physiol Biochem 48:909–930

Gu X, Gao Z, Yan Y, Wang X, Qiao Y, Chen Y (2017) *RdreB1BI* enhances drought tolerance by activating AQP-related genes in transgenic strawberry. Plant Physiol Biochem 119:33–42

Guy CL (1990) Cold acclimation and freezing stress tolerance: role of protein metabolism. Annu Rev Plant Physiol Plant Mol Biol 41:187–223

Hao J, Yang J, Dong J, Fei S (2017) Characterization of *BdCBF* genes and genome-wide transcriptome profiling of *BdCBF3*-dependent and -independent cold stress responses in *Brachypodium distachyon*. Plant Sci 262:52–61

Hare PD, Cress WA (1997) Metabolic implications of stress-induced proline accumulation in plants. Plant Growth Regul 21:79–102

Hu H, Dai M, Yao J, Xiao B, Li X, Zhang Q, Xiong L (2006) Overexpression a NAM, ATAF, and CUC(NAC) transcription factor enhances drought resistance and salt tolerance in rice. Proc Natl Acad Sci USA 103:12987–12992

Jangra S, Mishra A, Kamboj D, Yadav NR, Yadav RC (2017) Engineering abiotic stress tolerance traits for mitigating climate change. In: Gahlawat S, Salar R, Siwach P, Duhan J, Kumar S, Kaur P (eds) Plant biotechnology: recent advancements and developments. Springer, Singapore

Ji SJ, Lu YC, Feng JX, Wei G, Li J, Shi YH, Fu Q, Liu D, Luo JC, Zhu XY (2003) Isolation and analyses of genes preferentially expressed during early cotton fiber development by subtractive PCR and cDNA array. Nucleic Acids Res 31:2534–2543

Jin R, Kim HB, Ji CY, Kim HS, Li HM, Ma DF, Kwak SS (2017) Overexpressing *IbCBF3* increases low temperature and drought stress tolerance in transgenic sweet potato. Plant Physiol Biochem 118:45–54

Karaba A, Dixit S, Greco R, Aharonia A, Trijamtmiko KR, Marsch-Martinez N, Krishnan A, Nataraja KN, Udayakumar M, Pereira A (2007) Improvement of water use efficiency in rice expression of HARDY, an Arabidopsis drought and salt tolerance gene. Proc Natl Acad Sci USA 104:15270–15275

Khodarahmpour Z (2011) Genetic analysis of tolerance to heat stress in maize (Zea mays L). Afri J Agric Res 6:2767–2773

Kumar V, Shriram V, Hoque TS, Hasan MM, Burritt DJ, Hossain MA (2017) Glycinebetaine-mediated abiotic oxidative-stress tolerance in plants: physiological and biochemical mechanisms. In: Sarwat M, Ahmad A, Abdin M, Ibrahim M (eds) Stress signaling in plants: genomics and proteomics perspective, vol 2. Springer, Cham

Legrand S, Marque G, Blassiau C, Bluteau A, Canoy AS, Fontaine V, Jaminon O, Mautord J, Morin J, Petit A, Baranger A, Riviere N, Wilmer J, Delbreil B, Henaut IL (2013) Combining gene expression and genetic analyses to identify candidate genes involved in cold responses in pea. J Plant Physiol 170:1148–1157

Li H, Ding Y, Shi Y, Zhang X, Zhang S, Gong Z, Yang S (2017a) MPK3- and MPK6-mediated ICE1 phosphorylation negatively regulates ICE1 stability and freezing tolerance in *Arabidopsis*. Dev Cell 43:630–642

Li NN, Qian WJ, Wang L, Cao HL, Hao XY, Yang YJ, Wang XC (2017b) Isolation and expression features of hexose kinase genes under various abiotic stresses in the tea plant (*Camellia sinensis*). J Plant Physiol 209:95–104

Li WI, Sun Q, Li WC, Yu YL, Zhao M, Meng ZD (2017c) Characterization and expression analysis of a novel RING-HC gene, *ZmRHCP1*, involved in brace root development and abiotic stress responses in maize. J Integr Agr 16:1892–1899

Liu L, Song J, Xu J, Li D, Li G, An L (2017) Differential expression by chromatin modifications of alcohol dehydrogenase 1 of *Chorisporabungeana* in cold stress. Gene 636:1–16

Liu Q, Kasuga M, Sakuma Y, Abe H, Miura S, Yamaguchi-Shinozaki K, Shinozaki K (1998) Two transcription factors, DREB1 nad DREB2, with an EREBP/AP2DNA binding domain separate two cellular signal transduction pathways in drought and low temperature responsive gene expression, respectively, in Arabidopsis. Plant Cell 10:1391–1406

Ma X, Wang G, Zhao W, Yang M, Ma N, Kong F, Dong X, Meng Q (2017) SlCOR413IM1: a novel cold-regulation gene from tomato, enhances drought stress tolerance in tobacco. J Plant Physiol 216:88–99

Makarevitch I, Waters AJ, West PT, Stitzer M, Hirsch CN, Ibarra JR, Springer NM (2015) Transposable elements contribute to activation of maize genes in response to abiotic stress. PLoS Genet 11:e1004915

Masuka B, Araus JL, Das B, Sonder K, Cairns JE (2012) Phenotyping for abiotic stress tolerance in maize. J Integr Plant Biol 54:238–249
Melkonian J, Yu XL, Setter TL (2004) Chiling responses of maize (Zea mays L.) seedlings: root hydraulic conductance, abscisic acid, and stomatal conductance. J Exp Bot 55(403):1751–1760
Meseka SK, Menkir A, Ajala S (2011) Genetic analysis of performance of maize inbred lines under drought stress. J Crop Improv 25:521–539
Mickelbart MV, Hasegawa PM, Serres JB (2015) Genetic mechanism of abiotic stress tolerance that translate to crop yield stability. Nat Rev Genet 16:237–251
Mittler R (2002) Oxidative stress, antioxidants and stress tolerance. Trends Plant Sci 7:405–410
Mittler R (2006) Abiotic stress, the field environment and stress combination. Trends Plant Sci 11:15–19
Moalem SK, Amiri RM, SS K-S (2017) Effect of cold stress on oxidative damage and mitochondrial respiratory properties in chickpea. Plant Physiol Biochem 122:31–39
Nishiyama I (1995) Damage due to extreme temperatures. In: Matsuo T, Kumazawa K, IshiiR IH, Hirata H (eds) Science of the Rice Plant. Food and Agriculture Policy Research Center, Tokyo, pp 769–812
Okono A, Monneveux P, Ribaut JM (2014) Facing the challeneges of global agricultural today: what can we do about drought? Front Physiol 4:6–7
Qi XN, Xiao XY, Fan ZQ, Chen JY, Lu WJ, Kuang JF (2017) A banana fruit transcriptional repressor MaERF10 interacts with MaJAZ3 to strengthen the repression of JA biosynthetic genes involved in MeJA-mediated cold tolerance. Postharvest Biol Technol 120:222–231
Rao RSP, Anderson JR, Dionisio G, Boelt B (2017) Fructan accumulation and transcription of candidate genes during cold acclimation in three varieties of *Poapratensis*. J Plant Physiol 168:344–351
Rathiasabapathi B (2000) Metabolic engineering for stress tolerance: installing osmoprotectant synthesis pathways. Ann Bot 86:709–716
Richards RA, Rebetzke GJ, Watt M, Condon AG, Spielmeyer W, Dolferus R (2010) Breeding for improved water productivity in temperate cereals: phenotyping, quantitative trait loci, markers and the selection environment. Funct Plant Biol 37:85–97
Rihan HZ, Issawi M, Fuller MP (2017) Upregulation of *CBF/DREB1* and cold tolerance in artificial seeds of cauliflower (*Brassica oleracea* var. *botrytis*). Sci Hort 225:299–309
Rodriguez VM, Butron A, Rady MOA, Soengas P, Revilla P (2014) Identification of quantitvate trait loci involved in the response to cold stress in maize (Zea mays L.). Mol Breed 33:363–371
Salekdeh GH, Reynolds M, Bennett J, Boyer J (2009) Conceptual framework for drought phenotyping during molecular breeding. Trends Plant Sci 14:488–496
Schluter U, Colmsee C, Scholz U, Brautigam A, Weber APM, Zellerhoff N, Bucher M, Fahnenstich H, Sonnewald U (2013) Adaptation of maize source leaf metabolism to stress related disturbances in carbon, nitrogen and phosphorus balance. BMC Genomics 14:442–467
Seki M, Kamei A, Yamaguchi-Shinozaki K, Shinozaki K (2003) Molecular responses to drought, salinity and frost: common and different paths for plant protection. Curr Opin Biotechnol 14:194–199
Shen Y, Wu X, Liu D, Song S, Liu D, Wang H (2017) Cold-dependent alternative splicing of a Jumonji C domain-containing gene *MtJMJC5* in *Medicagotruncatula*. Biochem Biophys Res Commun 474:271–276
Sowinski P, Maleszewski S (1989) Chilling sensivity in maize seedlings. Pt. 1 Growth and functioning of shoots and roots. Acta Physiol Plant 11:165–171
Staggenborg SA, Vanderlip RL (1996) Sorghum grain yield reductions caused by duration and timing of freezing temperatures. Agron J 88:473–477
Su Y, Liang W, Liu Z, Wang Y, Zhao Y, Ijaz B, Hua J (2017) Overexpression of *GhDof1* improved salt and cold tolerance and seed oil content in *Gossypiumhirsutum*. J Plant Physiol 218:222–234
Subbaiah CC, Sachs MM (2003) Molecular and cellular adaptations of maize to flooding stress. Ann Bot 91:119–127
Takeda S, Matsuoka M (2008) Genetic approaches to crop improvement: responding to environmental and population changes. Nature Rev Genet 9:444–457

Verheul MJ, Picatto C, Stamp P (1996) Growth and development of maize (Zeamays L.) seedlings under chilling conditions in the field. Eur J Agron 5:31–43

Wang Q, Pu Y, Yang D, Yin X, He Z, Yang Y, Yang Y (2017) Molecular cloning and characterization of the glutathione reductase gene from *Stipapurpurea*. BiochemBiophys Res Commun (In Press)

Wang W, Vinocur B, Altman A (2003) Plant responses to drought, salinity and extreme temperatures: towards genetic engineering for stress tolerance. Planta 218:1–14

Wani SH, Sandhu JS, Gosal SS (2008) Genetic engineering of crop plants for abiotic stress tolerance. In: Malik CP, Kaur B, Wadhwani C (eds) Advanced topics in plant biotechnology and plant biology. MD Publications, New Delhi, pp 149–183

Welcker C, Boussuge B, Benciveni C, Ribaut JM, Tardieu F (2007) Are source and sink strengths genetically linked in maize plants subjected to water deficit? A QTL study of the responses of leaf growth and anthesis-silking interval to water deficit. J Exp Bot 58:339–349

Wilhelm EP, Mullen RE, Keeling PL, Singletary GW (1998) Heat stress during grain filling in maize: effects on kernel growth and metabolism. Crop Sci 39:1733–1741

Xu R, Gao H, Zhang S, Liu P, Wang X, Hao Y (2017) Genome-wide identification and phylogenetic, comparative genomic, alternative splicing, and expression analyses of TCP genes in plants. Plant Gene 12:23–32

Yang GP, Ross DT, Kang WW, Brown PO, Weigel RJ (1999) Combining SSH and cDNA microarrays for rapid identification of differentially expressed genes. Nucl Acids Res 27:1517–1523

Yang H, Song J, Yue L, Mo X, Song J, Mo B (2017) Identification and expression profiling of *Oryza sativa* nucleotidyl transferase protein (NTP) genes under various stress conditions. Gene 628:93–102

Ying M, Kidou S (2017) Discovery of novel cold-induced *CISP* genes encoding small RNA-binding proteins related to cold adaptation in barley. Plant Sci 260:129–138

Ying S, Zhang DF, Li HY, Liu YH, Shi YS, Song YC, Wang TY, Li Y (2011) Cloning and characterization of a maize SnRK2 protein kinase gene confers enhanced salt tolerance in transgenic Arabidopsis. Plant Cell Rep 30:1683–1699

Zhang T, Che F, Zhang H, Pan Y, Xu MQ, Ban Q, Han Y, Rao J (2017) Effect of nitric oxide treatment on chilling injury, antioxidant enzymes and expression of the CmCBF1 and CmCBF3 genes in cold-stored Hami melon (*Cucumismelo* L.) fruit. Postharvest Biol Technol 127:88–98

Zhang Y, Yu H, Yang X, Li Q, Ling J, Wang H, Gu X, Huang S, Jiang W (2016) *CsWRKY46*, a WRKY transcription factor from cucumber, confers cold resistance in transgenic-plant by regulating a set of cold-stress responsive genes in an ABA-dependent manner. Plant Physiol Biochem 108:478–487

Zheng J, Fu J, Gou M, Huai J, Liu Y, Jian M, Huang Q, Guo X, Dong Z, Wang H, Wang G (2010) Genome-wide transcriptome analysis of two maize inbred lines under drought stress. Plant Mol Biol 72:407–421

Zheng J, Zhao J, Tao Y, Wang J, Liu Y, Fu J, Jin Y, Gao P, Zhang J, Bai Y, Wang G (2004) Isolation and analysis of water stress induced genes in maize seedling by substractive PCR and cDNA microarray. Plant Mol Biol 55:807–823

Zhou A, Sun H, Feng S, Zhou M, Gong S, Wang J, Zhang S (2017) A novel cold-regulated gene from *Phlox subulata*, *PsCor413im1*, enhances low temperature tolerance in Arabidopsis. Biochem Biophys Res Commun (In Press)

Zou Y, Ma L, Dong H, Tao F, Feng X, Yuan D, Fan D, Hu X (2017) PstPS1, the trehalose-6-phosphate synthase gene of Puccinia striiformis f. sp. tritici, involves in cold stress response and hyphae development. Physiol Mol Plant Pathol 100: 201–208

Chapter 3
Redox Regulation of Cold Stress Response

Venura Herath

3.1 Introduction

Abiotic stress conditions are affecting every crop cultivated around the globe at various scales (Zandalinas et al. 2018). Unlike animals, the impact of these stresses is more severe in plants since they are sessile organisms. Throughout the evolution, plants developed various mechanisms to allure these stress conditions at molecular, cellular, and physiological levels. Identification and characterization of these complex mechanisms will hold the key to developing enhanced crops with abiotic stress tolerance. In order to facilitate the task, it is important to look at the stress response mechanisms starting from the perception of the signal, early response mechanisms, and subsequent series of secondary responses followed by the whole plant-level responses (Park et al. 2015). Most (might be all) of the instances, there is a cross talk between different abiotic stress response pathways (Sharma et al. 2013; Nakashima et al. 2014). This opens up an avenue for breeders toward developing crops with wide spectrum of tolerance to various abiotic stresses.

Cold stress is considered as one of the major abiotic stresses that contribute to yield losses worldwide. It can occur at high altitudes of tropical regions as well as during fall and winter seasons in semi-temperate regions, and therefore, it is not restricted to temperate regions. The majority of the major crops are susceptible to cold stress. Even though the frequency of occurrence of low-temperature events is decreasing due to climate change, cold stress is still causing significant yield loss around the world (Papagiannaki et al. 2014; Vitasse and Rebetez 2018). This highlights the importance of breeding crops that can withstand extreme temperatures. There are promising attempts to breed future crops using molecular breeding, modern genomic techniques, and combination of both approaches (Jha et al. 2017).

V. Herath (✉)
Department of Agricultural Biology, Faculty of Agriculture,
University of Peradeniya, Peradeniya, Sri Lanka
e-mail: venurah@pdn.ac.lk

S. H. Wani, V. Herath (eds.), *Cold Tolerance in Plants*,
https://doi.org/10.1007/978-3-030-01415-5_3

Among different cold stress regulatory strategies developed by plants, the redox regulatory mechanism is considered as one of the major mechanisms. Naturally, plants are adapted to high oxygen concentrations as a result of the photosynthesis process. Oxygen metabolism generates various reactive oxygen species (ROS). These molecules are produced in various organelles such as chloroplast, mitochondria, and peroxisome and in cytoplasm. ROS include singlet oxygen ($^1O_2$), hydroxyl radical ($OH^•$), superoxide radical ($O_2^{•-}$), and hydrogen peroxide (H_2O_2). Concentrations of these ROS largely vary within various cells as well as inside cells (Einset et al. 2007; Das et al. 2015). Higher concentrations of ROS have negative impacts on plant metabolism. When the accumulation of ROS in cells exceeds the threshold, cells will undergo oxidative stress which can ultimately result in death. Cold stress interferes with the CO_2 fixation process and limits $NADP^+$ generation via the Calvin-Benson-Bassham cycle. It results in over-reduction of electron transport chain. As a consequence, in chloroplasts, superoxide radicals and singlet oxygen molecules are generated. Plants deal with the situation by regenerating $NADP^+$ with the help of photorespiration process. This process results in the production of H_2O_2 in peroxisomes (Shao et al. 2006; Willems et al. 2016). However, specific concentrations of ROS are essential for cellular signaling processes involved in plant development and stress response. Due to this dual nature, ROS are considered as a double-edged sword. Therefore, plants maintain redox homeostasis with the help of various enzymatic and nonenzymatic antioxidant mechanisms. The major enzymatic antioxidants include catalase (CAT), peroxidases (POX), superoxide dismutase (SOD), and glutathione reductase (GR), while ascorbic acid (AA), α-tocopherol, glutathione, and carotenoids act as nonenzymatic antioxidants (Choudhury et al. 2013; Gupta et al. 2018). Efficient antioxidant systems that neutralize the oxidative stress conditions enable plants to withstand various stress conditions (Farooq et al. 2009a; Türkan and Demiral 2009). Bioengineering has been effectively used to express multiple antioxidant enzymes and compounds that lead to develop tolerant plants to multiple abiotic and also biotic stress conditions (Sahoo et al. 2013; Chakraborty and Bhattacharjee 2015).This chapter is focusing on components, behavior, interplay, and regulation of redox regulation mechanism under cold stress and strategies for the production of oxidative stress-tolerant plants that can withstand cold and other stress conditions.

3.2 ROS: Production Under Cold Stress

ROS generation occurs as a result of oxygen metabolism and stress conditions. Generation of ROS as a result of basal oxidative metabolism rarely reaches oxidative stress levels in plant cells. However, a stress or multiple stress conditions induce ROS production and accumulation of ROS at high levels. Chloroplasts, mitochondria, and peroxisomes act as the major sites for the production of ROS. In chloroplasts, leakage of electrons happens from Fe-S centers of photosystem I (PSI) and as a result of reduction of ferredoxin. Consequently, $O_2^{•-}$ is generated. $O_2^{•-}$ is then converted to H_2O_2 by SOD. $^1O_2$ can be generated by PSII. Stress conditions interfere with the CO_2 fixation in the chloroplasts resulting in enhanced oxygenase activity of Rubisco. This initiates photorespiration process. Glycolate is generated as a result, and then it is transferred to peroxisomes. Inside peroxisome, glycolate is converted to glyoxylate with the help of glycolate oxidase. This reaction leads to the formation of H_2O_2. H_2O_2 is produced by xanthine oxidase using O_2 as the substrate. Inside mitochondria, ROS production is mainly localized to complex I and III. Cell wall-associated peroxidases and plasma membrane-bound NADPH oxidases are also involved in generating $O_2^{\cdot-}$ and H_2O_2. Excess accumulation of ROS causes damages to membranes, nucleic acids, proteins, etc. resulting in the accumulation of toxic compounds in cytoplasm, metabolic imbalances, and eventually cell death (Hossain et al. 2015).

Cold stress is collectively used to identify chilling stress (<10 °C) and freezing stress/frost (<0 °C). Survival of plants under cold stress depends on perception and rapid initiation of cold stress response signal transduction networks (Yun et al. 2010; Park et al. 2010). Perception of cold signal by unknown cell wall or membrane sensors probably due to alteration of fluidity of membranes activates calcium channels that facilitate the influx of Ca^{2+} ions. This influx will trigger a cascade of secondary messengers such as H_2O_2, NO, and ABA. It has been shown that cold stress induces the expression and activity of ROS scavengers (Heidarvand and Maali Amiri 2010; Janská et al. 2010).The responses can vary with different species (Gechev et al. 2006). Cold stress creates an imbalance between absorption and utilization of light by interfering with Calvin-Benson-Bassham cycle. It also causes over-reduction of electron transport chain reactions of respiration process and increasing photosynthetic electron flux to O_2 (Logan et al. 2006; Hu et al. 2008).

3.3 Oxidative Damage Induced by Cold Stress

Crops grown around the world are affected by low-temperature conditions. Chilling stress is considered as one of the devastating abiotic stress conditions that critically affects growth, development, and reproduction. It induces tissue dehydration also known as physiological drought. It leads to the accumulation of malondialdehyde through lipid peroxidation that eventually leads to membrane disintegration. Extensive membrane damage hinders the plant growth and development causing extensive yield losses in corn (Farooq et al. 2009b; Hasanuzzaman et al. 2013). In wheat, exposure to 3 °C caused reduction in chlorophyll content, assimilation of CO_2, and reduction of transpiration rate and interfered with photosynthesis. These conditions were a result of reduction in ATP synthase function. Legume crops also show sensitivity to chilling temperatures. Initially, chilling stress accelerates the respiratory rate followed by the reduction of respiratory rate that results in the disruption of respiration. Chilling stress also interferes with photosynthesis in legumes like mung bean, chickpea and pigeon pea, and soybean by altering chloroplast development and photosystem II and inducing photo inhibition process

(Srougi et al. 2013; Hetherington et al. 1989; Yang et al. 2005; Turan and Ekmekçi 2011). Cold-induced oxidative stress triggers the activation of antioxidant enzymes like CAT, POX, SOD, and GR. However, antioxidant mechanism tends to fail after prolong exposure or increase intensity of cold stress by the plant. As a result, ROS accumulation increases resulting in severe damage to cellular components and homeostasis. Cold stress-tolerant varieties are shown to be containing effective antioxidant mechanisms and capacity (del Río et al. 2018).

3.4 Redox Signaling

In addition to its cytotoxic role, ROS can act as secondary messengers. ROS signaling is involved in myriad of signaling networks important for plant growth and stress tolerance (Sies et al. 2017; Kundu et al. 2018). Among different ROS, $O_2^{\cdot-}$ and H_2O_2 act as major signaling molecules (Petrov and Van Breusegem 2012; Choudhury et al. 2013). Both the generation and changes in concentration of ROS will trigger various unique and shared gene expression networks in the nucleus. Due to ROS's short half-life, it is very difficult to decode ROS signal transduction pathways. As a result, fine regulatory mechanisms of ROS signaling are yet to be discovered even though there is significant progress made during the last decade (Sies et al. 2017). Both field experiments and simulation experiments were used for the identification of ROS signaling mechanism. As an example, rice seedlings exposed to chilling stress (10 °C) have been shown to accumulate ROS that leads to the activation of a signaling cascade controlled mainly via OsTGA10 transcription factor. OsTGA10 regulon can also be triggered by exogenous application of H_2O_2-simulating chilling-induced signal transduction network (Yun et al. 2010; Herath 2011).

Recent advances point to an unknown histidine kinase-like ROS receptor for sensing both intracellular and extracellular H_2O_2 (Baxter et al. 2014). Upon ROS sensing, ROS receptor induces the generation of phosphatidic acid and triggers Ca^{2+} release from both cell wall and vacuole (Fig. 3.1). This induces the activation of kinases heterotrimeric G-proteins and other transcription factors. Mitogen-activated protein kinase (MAPK) cascade has been shown to activate ROS signaling transcription factors through phosphorylation-activation process of a MAP kinase kinase kinase (MAPKKK), a MAP kinase kinase (MAPKK), and a MAP kinase (MAPK) (Zhang et al. 2018). Alternatively, intercellular H_2O_2 also has the ability to activate downstream transcription factors (Petrov and Van Breusegem 2012; Petrov et al. 2015). Ca^{2+} also has been shown to interact with ROS signaling process. Release of Ca^{2+}can be triggered by ROS, ABA, or a combination of both ROS and ABA. It directly activates MAPK cascade. MEKK1-MKK1/2-MPK4/6 signaling cascade triggered by cold stress has shown to display a two-way communication with ROS. MEKK1 has been reported to be stimulated and stabilized by H_2O_2. MPK4 and MPK6 have been found to be activated by ROS (Teige et al. 2004).

The activation of ROS-responsive transcription factors triggers downstream defense- related genes. These genes share common *cis*-element signatures.

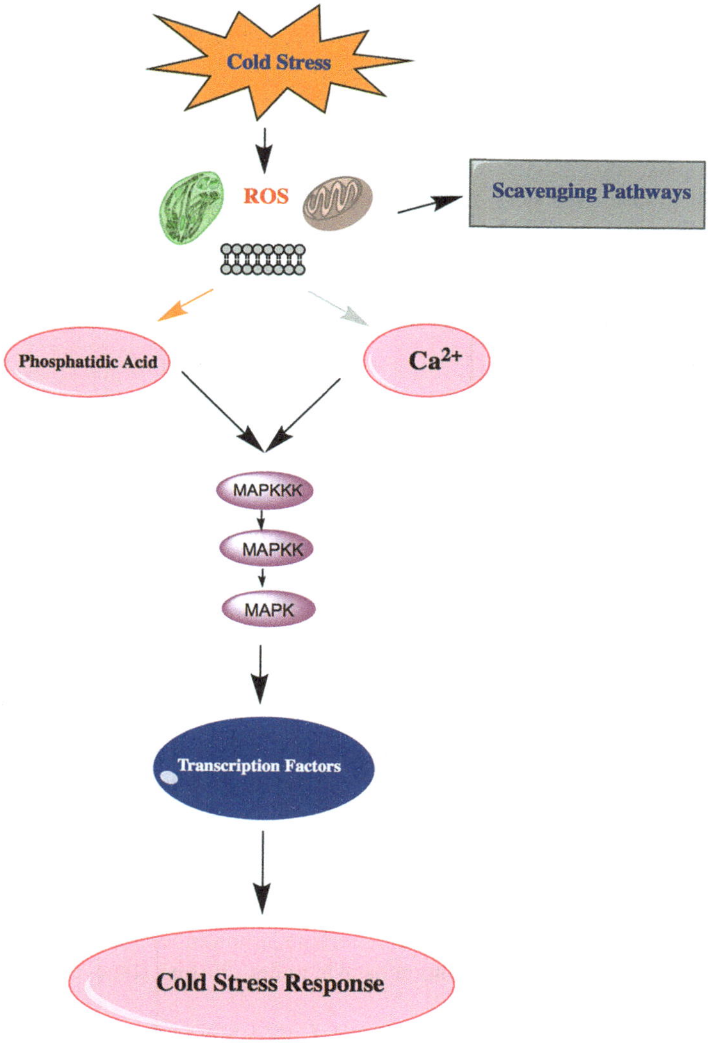

Fig. 3.1 Signaling pathways involved in the perception and transduction of ROS under cold stress conditions

Among such cis-elements, as1/ocs/TGA-like *cis*-elements (consensus sequences; “AATTTGAT,” “TAATTTGA”) and TGA core motif are considered as major elements responsible for ROS signaling in *Arabidopsis*, tobacco, and rice (Garretón et al. 2002; Yun et al. 2010). Plant growth regulators such as abscisic acid, salicylic acid, ethylene, and jasmonic acid have been shown to act downstream after induction by ROS. However, ROS is also exhibited to act as a secondary messenger in hormonal signaling networks. This shows the complex nature of ROS signaling in plants (Mühlenbock et al. 2008; Sharma et al. 2013; Cheng et al. 2016; Souza et al. 2017).

Proper communication between cellular organelles is essential for survival under oxidative stress conditions. Retrograde signaling related to ROS generation is sent to the nucleus by two major sites of ROS production: chloroplast and mitochondria. These signals play a major role in stress tolerance. Mg-protoporphyrin IX (Mg-PPIX) is one of the most studied stress-responsive retrograde signaling pathways between the chloroplast and nucleus. It was shown that ~35% of proteins including glutathione S-transferases and peroxidases identified in Mg-PPIX are involved in stress response (Suzuki et al. 2012; Choudhury et al. 2013). Unlike chloroplast retrograde signaling, details of mitochondrial retrograde signaling are yet to be discovered. There is some evidence showing alternative oxidase 1 (*AOX1*) as one of the important components in mitochondrial retrograde signaling (Saika et al. 2002; Fiorani 2005; Polidoros et al. 2005).

3.5 Redox Homeostasis

Accumulation of excessive amount of ROS can create threatening conditions for plants. In order to prevent such damage, plants enhance expression of antioxidant enzymes and molecules. Plants with enhanced expression of antioxidants have shown to be more tolerant to stress than the ones without such. As an example, ectopic overexpression of APX of pea in tomato enhances the tolerance not only to chilling but also for salt stress indicating the importance of antioxidative enzymes in stress response (Wang et al. 2005). .

Plant cells generate high concentrations of ascorbate which act as a hydrophilic redox buffer. It provides strong defense against oxidative stress conditions. Also, thols in reduced state acting as thol buffer is also important in redox regulation. Redox balance is maintained by these antioxidants through regulation of the level of reductants and oxidants in a balanced state. Tocopherols such as vitamin E also act as liposoluble redox buffers. Tocopherols are involved in singlet oxygen scavenging as well as other ROS. Collectively, ascorbate, glutathione, and tocopherols are essential for maintaining redox homeostasis under chilling stress conditions (Sies et al. 2017; Gupta et al. 2018).

Scavenging of $O_2^{\cdot-}$ into H_2O_2 by Fe-SOD, Cu/Zn-SOD, and Mn-SOD superoxide dismutase isoforms plays a major role in redox homeostasis. These enzymes are present in all cellular compartments. Resulting H_2O_2 is catalyzed by peroxisomal catalase through ascorbate-glutathione (ASC-GSH) cycle and thioredoxin/peroxiredoxin (Trx/Prx) system. In ASC-GSH cycle, H_2O_2 scavenging occurs via APX with the help of reduced ASC. As a result, ASC will be oxidized to monodehydroascorbate and dehydroascorbate. Then they will be reduced by the enzyme FAD-containing monodehydroascorbate reductase and dehydroascorbate reductase (DHAR) in the presence of NADPH and GSH (Yamamoto and Nasrallah 2013; Sies et al. 2017).

Thioredoxins, peroxiredoxin, and sulfiredoxin are the major redox protein constitutes in Trx/Prx system. They are involved in the reduction of thol residues of oxidized proteins found in various cellular locations. Formation of reversible disulfide bonds is a protective mechanism against overoxidation of Cys residues under optimal and also under slightly oxidant conditions. Overoxidation of Cys residues leads to protein degradation. Trxs and glutaredoxins act as main systems regulating thiol-disulfide groups in proteins. They also can act as substrates of peroxiredoxins (Prxs) or ribonucleotide reductases. Oxidization/reduction rates of GSH and Trx are mainly responsible for the redox state of various cellular localities (Poynton and Hampton 2014; Sevilla et al. 2015; Gupta et al. 2018).

At organelle level, redox status is finely regulated. In chloroplasts, changes in redox status induce the expression of proteins such as plastoquinone, ascorbate, GHS, and components of Trx/Prx system. Catalase, APX, and ASC-GSH system are involved in the scavenging of H_2O_2 in peroxisomes. Even though the production of ROS is relatively low in plant mitochondria, significant oxidation of electron transport chain proteins was observed especially in electron transport complex I and III. Reduced level of ROS production is considered to be a result of the presence of alternative oxidases (Maxwell et al. 1999; Dahal and Vanlerberghe 2017). Programmed cell death is induced when the ROS levels in mitochondria exceed the threshold especially under various stressful conditions (Petrov et al. 2015).

3.6 Role of ROS in Plant Acclimation

As sessile organisms, plants show remarkable ability to acclimate to various environment conditions both rapid and evolutionary manner. This is facilitated by changes in biochemical, physiological, anatomical, and genetic levels. Plants show broad spectrum of responses to cold stress conditions. Some plants can withstand sub-zero temperatures after acclimation (freeze tolerant), while some plants can tolerate low but above zero temperatures (chilling tolerant). There is a group of plants that do not have the ability to survive under chilling temperatures (chilling susceptible). These responses are mainly governed by their center of origin and various evolutionary stages that they underwent in the past. Cold acclimation is achieved via prior exposure to low and nonfreezing, suboptimal temperatures (Jiang et al. 2013; Shi et al. 2015). As a result of acclimation process, plant starts synthesis of cryoprotectants, dehydrin proteins, cold-regulated proteins and heat-shock proteins, and various antioxidants. Cryoprotective compounds include low-molecular-weight nitrogenous compounds (proline, glycine betaine), soluble sugars (raffinose, trehalose), and sugar alcohols (sorbitol, mannitol). They play a major role in membrane stabilization under cold stress conditions since the membrane is the main target of cold stress. These compounds interact closely with dehydrin proteins, cold-regulated proteins, and heat-shock proteins stabilizing membrane phospholipids and proteins. They also play a role in ion homeostasis, ice nucleation, and ROS scavenging. Both antioxidative enzymes and nonenzymatic antioxidants also play a major role in cold acclimation. Enhanced activity of these antioxidants facilitates redox homeostasis and cold stress conditions. In addition, membrane lipid composition is also altered by increasing the proportion of unsaturated fatty acids in order to maintain the membrane functionality under stress (Lynch and Steponkus 1987; Welti et al. 2002; van Meer et al. 2008).

3.7 Taming Redox Regulation for the Development of Stress- Tolerant Plants

Transgenic approaches have shown that enhanced abiotic stress response can be achieved by enhancing various players of redox regulation mechanism (Table 3.1). Signal-independent overexpression of endogenous antioxidative enzymes as well as ectopic expression was successfully used to engineer plants with higher tolerance to cold stress. Overexpression of APX enhances the cold tolerance in rice at booting stage (Sato et al. 2011). Ectopic expression of wheat CAT in rice results in controlled H_2O_2 levels up to 8 days under 5 °C (Matsumura et al. 2002). Nonenzymatic antioxidants have also been identified as potential targets to alleviate oxidative stress under stress conditions. Overexpression of Δ 1-pyrroline-5-carboxylate synthetase is also shown to enhance stress tolerance (Vendruscolo et al. 2007). Overexpression of transcription factors involved in oxidative stress response exhibits enhanced cold stress tolerance. In rice, overexpression of *OsTGA10* and *OsMYB4*enhances the chilling stress response at seedling stage under 10 °C (Park et al. 2010; Herath 2011). Co-expression of genes involved in redox regulation has also shown promise over transgenics carrying only one gene. Overexpression of MeAPX2 together with MeCu/ZnSOD results in enhanced oxidative and chilling response in cassava (Xu et al. 2014)

Table 3.1 Development of transgenics for cold stress tolerance by manipulating oxidative stress signaling

Gene/s	Host plant	Function	Reference
Superoxide dismutase (SOD)	*Nicotiana tabacum*	ROS scavenging	Gupta et al. (1993)
Glutathione-S-transferase/ glutathione peroxidase (gst/gpx)	*Nicotiana tabacum*	ROS scavenging	Roxas et al. (1997)
Glutathione peroxidase (GPX-2)	*Arabidopsis thaliana*	ROS scavenging	Gaber et al. (2006)
Catalase (CAT)	*Oryza sativa*	ROS scavenging	Matsumura et al. (2002)
CRT/DRE element binding factor (*CBF1*)	*Solanum lycopersicum*	Enhanced SOD production	Zhang et al. (2013)
Dehydrins (*RAB18* and *COR47* or *LTI29* and *LTI30*)	*Arabidopsis thaliana*	Membrane stabilization	Puhakainen et al. (2004)
S-adenosylmethionine decarboxylase (*CdSAMDC1*)	*Eremochloa ophiuroides*	Regulation of H_2O_2 generation	Chinnusamy et al. (2017)

3.8 Conclusions and Future Prospects

Cold stress is considered as a key abiotic stress that affects crop production worldwide. ROS play a major role in both oxidative stress and oxidative signaling under cold stress conditions. Accumulation of ROS leads to membrane, lipid, nucleic acid, and protein damage. Plants have developed various redox regulation mechanisms to combat these suboptimal levels of ROS. These mechanisms include ROS scavenging powered by antioxidative enzymes and antioxidant molecules. In addition to their cytotoxic nature, ROS are playing a major role as signaling molecules involved in various developmental, reproductive, and stress response signal transduction networks. ROS signaling is also closely associated with various plant hormonal signaling networks. Even though there is a myriad of information available, our knowledge about synthesis, regulation, and signaling ROS is still not adequate. Future research is needed to answer open questions about redox regulation in plants. With the advancement of novel genomic editing platforms and next-generation sequencing technologies, it will facilitate the further exploration of oxidative stress signaling and ultimately the production of cold- and other stress-tolerant crops.

Acknowledgment This work was supported by the National Research Council of Sri Lanka (NRC/14/117) and International Science Foundation, Sweden (C/5267-1).

References

Baxter A, Mittler R, Suzuki N (2014) ROS as key players in plant stress signalling. J Exp Bot 65:1229–1240. https://doi.org/10.1093/jxb/ert375

Chakraborty A, Bhattacharjee S (2015) Differential competence of redox-regulatory mechanism under extremes of temperature determines growth performances and cross tolerance in two indica rice cultivars. J Plant Physiol 176:65–77. https://doi.org/10.1016/j.jplph.2014.10.016

Cheng F, Lu J, Gao M et al (2016) Redox signaling and CBF-responsive pathway are involved in salicylic acid-improved photosynthesis and growth under chilling stress in watermelon. Front Plant Sci 7:1–16. https://doi.org/10.3389/fpls.2016.01519

Chinnusamy V, Hasanuzzaman M, Guo Z et al (2017) Transgenic centipedegrass (Eremochloa ophiuroides [Munro] Hack.) overexpressing S-adenosylmethionine decarboxylase (SAMDC) gene for improved cold tolerance through involvement of H 2 O 2 and NO signaling. Front Plant Sci 8:1655. https://doi.org/10.3389/fpls.2017.01655

Choudhury S, Panda P, Sahoo L, Panda SK (2013) Reactive oxygen species signaling in plants under abiotic stress. Plant Signal Behav 8:e23681. https://doi.org/10.4161/psb.23681

Dahal K, Vanlerberghe GC (2017) Alternative oxidase respiration maintains both mitochondrial and chloroplast function during drought. New Phytol 213:560–571. https://doi.org/10.1111/nph.14169

Das P, Nutan KK, Singla-Pareek SL, Pareek A (2015) Oxidative environment and redox homeostasis in plants: dissecting out significant contribution of major cellular organelles. Front Environ Sci 2:1–11. https://doi.org/10.3389/fenvs.2014.00070

del Río LA, Corpas FJ, López-Huertas E, Palma JM (2018) Plant superoxide dismutases: function under abiotic stress conditions. Antioxidants Antioxid Enzym High Plants:1–26. https://doi.org/10.1007/978-3-319-75088-0_1

Einset J, Winge P, Bones A (2007) ROS signaling pathways in chilling stress. Plant Signal Behav 2:365–367. https://doi.org/10.4161/psb.2.5.4461

Farooq M, Wahid A, Kobayashi N et al (2009a) Plant drought stress: effects, mechanisms and management. Sustain Agric 29:185–212. https://doi.org/10.1051/agro

Farooq M, Wahid A, Lee D-J et al (2009b) Advances in drought resistance of rice. CRC Crit Rev Plant Sci 28:199–217. https://doi.org/10.1080/07352680902952173

Fiorani F (2005) The alternative oxidase of plant mitochondria is involved in the acclimation of shoot growth at low temperature. A study of Arabidopsis AOX1a transgenic plants. Plant Physiol 139:1795–1805. https://doi.org/10.1104/pp.105.070789

Gaber A, Yoshimura K, Yamamoto T et al (2006) Glutathione peroxidase-like protein of Synechocystis PCC 6803 confers tolerance to oxidative and environmental stresses in transgenic Arabidopsis. Physiol Plant 128:251–262. https://doi.org/10.1111/j.1399-3054.2006.00730.x

Garretón V, Carpinelli J, Jordana X, Holuigue L (2002) The as-1 promoter element is an oxidative stress-responsive element and salicylic acid activates it via oxidative species. Plant Physiol 130:1516–1526. https://doi.org/10.1104/pp.009886

Gechev TS, Van Breusegem F, Stone JM et al (2006) Reactive oxygen species as signals that modulate plant stress responses and programmed cell death. BioEssays 28:1091–1101. https://doi.org/10.1002/bies.20493

Gupta AS, Webb RP, Holaday AS, Allen RD (1993) Overexpression of superoxide dismutase protects plants from oxidative stress (induction of ascorbate peroxidase in superoxide dismutase-overexpressing plants). Plant Physiol 103:1067–1073. https://doi.org/10.1104/pp.103.4.1067

Gupta DK, Palma JM, Corpas FJ (2018) Antioxidants and antioxidant enzymes in higher plants. Springer.

Hasanuzzaman M, Nahar K, Alam MM et al (2013) Physiological, biochemical, and molecular mechanisms of heat stress tolerance in plants. Int J Mol Sci 14:9643–9684. https://doi.org/10.3390/ijms14059643

Heidarvand L, Maali Amiri R (2010) What happens in plant molecular responses to cold stress? Acta Physiol Plant 32:419–431. https://doi.org/10.1007/s11738-009-0451-8

Herath V (2011) Transcriptional regulatory networks involved in plant responses to low temperature. The University of Maine, Orono.

Hetherington SE, He J, Smillie RM (1989) Photoinhibition at low temperature in chilling-sensitive and -resistant plants. Plant Physiol 90:1609–1615. https://doi.org/10.1104/pp.90.4.1609

Hossain MA, Bhattacharjee S, Armin S-M et al (2015) Hydrogen peroxide priming modulates abiotic oxidative stress tolerance: insights from ROS detoxification and scavenging. Front Plant Sci 6:1–19. https://doi.org/10.3389/fpls.2015.00420

Hu WH, Song XS, Shi K et al (2008) Changes in electron transport, superoxide dismutase and ascorbate peroxidase isoenzymes in chloroplasts and mitochondria of cucumber leaves as influenced by chilling. Photosynthetica 46:581–588. https://doi.org/10.1007/s11099-008-0098-5

Janská A, Maršík P, Zelenková S, Ovesná J (2010) Cold stress and acclimation - what is important for metabolic adjustment? Plant Biol 12:395–405. https://doi.org/10.1111/j.1438-8677.2009.00299.x

Jha UC, Bohra A, Jha R (2017) Breeding approaches and genomics technologies to increase crop yield under low-temperature stress. Plant Cell Rep 36:1–35. https://doi.org/10.1007/s00299-016-2073-0

Jiang Y, Peng D, Bai L-P et al (2013) Molecular switch for cold acclimation — anatomy of the cold-inducible promoter in plants. Biochemist 78:342–354. https://doi.org/10.1134/S0006297913040032

Kundu P, Gill R, Ahlawat S et al (2018) Targeting the redox regulatory mechanisms for abiotic stress tolerance in crops. In Biochemical, Physiological and Molecular Avenues for Combating Abiotic Stress Tolerance in Plants, pp. 151–220.

Logan BA, Kornyeyev D, Hardison J, Holaday AS (2006) The role of antioxidant enzymes in photoprotection. Photosynth Res 88:119–132. https://doi.org/10.1007/s11120-006-9043-2

Lynch DV, Steponkus PL (1987) Plasma membrane lipid alterations associated with cold acclimation of winter Rye seedlings (Secale cereale L. cv Puma). Plant Physiol 83:761–767. https://doi.org/10.1104/pp.83.4.761

Matsumura T, Tabayashi N, Kamagata Y et al (2002) Wheat catalase expressed in transgenic rice can improve tolerance against low temperature stress. Physiol Plant 116:317–327. https://doi.org/10.1034/j.1399-3054.2002.1160306.x
Maxwell DP, Wang Y, McIntosh L (1999) The alternative oxidase lowers mitochondrial reactive oxygen production in plant cells. Proc Natl Acad Sci 96:8271–8276. https://doi.org/10.1073/pnas.96.14.8271
Mühlenbock P, Szechynska-Hebda M, Plaszczyca M et al (2008) Chloroplast signaling and LESION SIMULATING DISEASE1 regulate crosstalk between light acclimation and immunity in Arabidopsis. Plant Cell 20:2339–2356. https://doi.org/10.1105/tpc.108.059618
Nakashima K, Yamaguchi-Shinozaki K, Shinozaki K (2014) The transcriptional regulatory network in the drought response and its crosstalk in abiotic stress responses including drought, cold, and heat. Front Plant Sci 5:170. https://doi.org/10.3389/fpls.2014.00170
Papagiannaki K, Lagouvardos K, Kotroni V, Papagiannakis G (2014) Agricultural losses related to frost events: use of the 850 hPa level temperature as an explanatory variable of the damage cost. Nat Hazards Earth Syst Sci 14:2375–2381. https://doi.org/10.5194/nhess-14-2375-2014
Park M-R, Yun K-Y, Mohanty B et al (2010) Supra-optimal expression of the cold-regulated OsMyb4 transcription factor in transgenic rice changes the complexity of transcriptional network with major effects on stress tolerance and panicle development. Plant Cell Environ 33:2209–2230. https://doi.org/10.1111/j.1365-3040.2010.02221.x
Park S, Lee C-M, Doherty CJ et al (2015) Regulation of the Arabidopsis CBF regulon by a complex low temperature regulatory network. Plant J 82:193. https://doi.org/10.1111/tpj.12796
Petrov V, Hille J, Mueller-Roeber B, Gechev TS (2015) ROS-mediated abiotic stress-induced programmed cell death in plants. Front Plant Sci 6:69. https://doi.org/10.3389/fpls.2015.00069
Petrov VD, Van Breusegem F (2012) Hydrogen peroxide--a central hub for information flow in plant cells. AoB Plants 2012:pls014. https://doi.org/10.1093/aobpla/pls014
Polidoros AN, Mylona PV, Pasentsis K et al (2005) The maize alternative oxidase 1a (Aox1a) gene is regulated by signals related to oxidative stress. Redox Rep 10:71–78. https://doi.org/10.1179/135100005X21688
Poynton RA, Hampton MB (2014) Peroxiredoxins as biomarkers of oxidative stress. Biochim Biophys Acta - Gen Subj 1840:906–912. https://doi.org/10.1016/j.bbagen.2013.08.001
Puhakainen T, Hess MW, Mäkelä P et al (2004) Overexpression of multiple dehydrin genes enhances tolerance to freezing stress in Arabidopsis. Plant Mol Biol 54:743–753. https://doi.org/10.1023/B:PLAN.0000040903.66496.a4
Roxas VP, Smith RK, Smith RK, Allen RD (1997) Overexpression of glutathione s-transferase/glutathione peroxidase enhances the growth of transgenic tobacco seedlings during stress. Nat Biotechnol 15:988–991. https://doi.org/10.1038/nbt1097-988
Sahoo KK, Tripathi AK, Pareek A, Singla-Pareek SL (2013) Taming drought stress in rice through genetic engineering of transcription factors and protein kinases. Plant Stress 7(1):60–72.
Saika H, Ohtsu K, Hamanaka S et al (2002) AOX1c, a novel rice gene for alternative oxidase; comparison with rice AOX1a and AOX1b. Genes Genet Syst 77:31–38. https://doi.org/10.1266/ggs.77.31
Sato Y, Masuta Y, Saito K et al (2011) Enhanced chilling tolerance at the booting stage in rice by transgenic overexpression of the ascorbate peroxidase gene, OsAPXa. Plant Cell Rep 30:399–406. https://doi.org/10.1007/s00299-010-0985-7
Sevilla F, Camejo D, Ortiz-Espín A et al (2015) The thioredoxin/peroxiredoxin/sulfiredoxin system: current overview on its redox function in plants and regulation by reactive oxygen and nitrogen species. J Exp Bot 66:2945–2955. https://doi.org/10.1093/jxb/erv146
Shao HB, Chu LY, Zhao CX et al (2006) Plant gene regulatory network system under abiotic stress. Acta Biol Szeged 50:1–9
Sharma R, De Vleesschauwer D, Sharma MK, Ronald PC (2013) Recent advances in dissecting stress-regulatory crosstalk in rice. Mol Plant 6:250–260. https://doi.org/10.1093/mp/sss147
Shi Y, Ding Y, Yang S (2015) Cold signal transduction and its interplay with phytohormones during cold acclimation. Plant Cell Physiol 56:7–15. https://doi.org/10.1093/pcp/pcu115
Sies H, Berndt C, Jones DP (2017) Oxidative stress. Annu Rev Biochem 86:715–748. https://doi.org/10.1146/annurev-biochem-061516-045037

Souza LA, Monteiro CC, Carvalho RF et al (2017) Dealing with abiotic stresses: an integrative view of how phytohormones control abiotic stress-induced oxidative stress. Theor Exp Plant Physiol 29:109–127. https://doi.org/10.1007/s40626-017-0088-8
Srougi MC, Miller HB, Witherow DS, Carson S (2013) Assessment of a novel group-centered testing schema in an upper-level undergraduate molecular biotechnology course. Biochem Mol Biol Educ 41:232–241. https://doi.org/10.1002/bmb.20701
Suzuki N, Koussevitzky S, Mittler R, Miller G (2012) ROS and redox signalling in the response of plants to abiotic stress. Plant Cell Environ 35:259–270. https://doi.org/10.1111/j.1365-3040.2011.02336.x
Teige M, Scheikl E, Eulgem T et al (2004) The MKK2 pathway mediates cold and salt stress signaling in Arabidopsis. Mol Cell 15:141–152. https://doi.org/10.1016/j.molcel.2004.06.023
Turan Ö, Ekmekçi Y (2011) Activities of photosystem II and antioxidant enzymes in chickpea (Cicer arietinum L.) cultivars exposed to chilling temperatures. Acta Physiol Plant 33:67–78. https://doi.org/10.1007/s11738-010-0517-7
Türkan I, Demiral T (2009) Recent developments in understanding salinity tolerance. Environ Exp Bot 67:2–9. https://doi.org/10.1016/j.envexpbot.2009.05.008
van Meer G, Voelker DR, Feigenson GW (2008) Membrane lipids: where they are and how they behave. Nat Rev Mol Cell Biol 9:112–124. https://doi.org/10.1038/nrm2330
Vendruscolo ECG, Schuster I, Pileggi M et al (2007) Stress-induced synthesis of proline confers tolerance to water deficit in transgenic wheat. J Plant Physiol 164:1367–1376. https://doi.org/10.1016/j.jplph.2007.05.001
Vitasse Y, Rebetez M (2018) Unprecedented risk of spring frost damage in Switzerland and Germany in 2017. Clim Chang 149:233–246. https://doi.org/10.1007/s10584-018-2234-y
Wang Y, Wisniewski M, Meilan R et al (2005) Overexpression of cytosolic ascorbate peroxidase in tomato confers tolerance to chilling and salt stress. J Am Soc Hortic Sci 130:167–173
Welti R, Li W, Li M et al (2002) Profiling membrane lipids in plant stress responses: role of phospholipase Dα in freezing-induced lipid changes in arabidopsis. J Biol Chem 277:31994–32002. https://doi.org/10.1074/jbc.M205375200
Willems P, Mhamdi A, Simon S et al (2016) The ROS wheel: refining ROS transcriptional footprints in Arabidopsis. Plant Physiol 171(2016):1720. https://doi.org/10.1104/pp.16.00420
Xu J, Yang J, Duan X et al (2014) Increased expression of native cytosolic Cu/Zn superoxide dismutase and ascorbate peroxidase improves tolerance to oxidative and chilling stresses in cassava (Manihot esculenta Crantz). BMC Plant Biol 14:208. https://doi.org/10.1186/s12870-014-0208-4
Yamamoto M, Nasrallah JB (2013) In planta assessment of the role of thioredoxin h proteins in the regulation of S-locus receptor kinase signaling in transgenic Arabidopsis. Plant Physiol 163:1387–1395. https://doi.org/10.1104/pp.113.225672
Yang MT, Chen SL, Lin CY, Chen YM (2005) Chilling stress suppresses chloroplast development and nuclear gene expression in leaves of mung bean seedlings. Planta 221:374–385. https://doi.org/10.1007/s00425-004-1451-y
Yun K-Y, Park MR, Mohanty B et al (2010) Transcriptional regulatory network triggered by oxidative signals configures the early response mechanisms of japonica rice to chilling stress. BMC Plant Biol 10:16. https://doi.org/10.1186/1471-2229-10-16
Zandalinas SI, Mittler R, Balfagón D et al (2018) Plant adaptations to the combination of drought and high temperatures. Physiol Plant 162:2–12. https://doi.org/10.1111/ppl.12540
Zhang L, Li Z, Li J, Wang A (2013) Ectopic overexpression of SsCBF1, a CRT/DRE-binding factor from the nightshade plant Solanum lycopersicoides, confers freezing and salt tolerance in transgenic Arabidopsis. PLoS One 8:15–17. https://doi.org/10.1371/journal.pone.0061810
Zhang M, Su J, Zhang Y et al (2018) Conveying endogenous and exogenous signals: MAPK cascades in plant growth and defense. Curr Opin Plant Biol 45:1–10. https://doi.org/10.1016/j.pbi.2018.04.012

Chapter 4
Hormonal Regulation of Cold Stress Response

Mohammad Arif Ashraf and Abidur Rahman

4.1 Introduction

Plant growth and development is a complex process mediated by individual and concerted action of phytohormones. From the seed germination to senescence, all the steps of development are the reflection of hormonal interplay. Not surprisingly, hormones also play major roles in promoting the adaptation of plants under abiotic stresses. The plant adaptation process during abiotic stress is mediated through the biosynthesis, transportation, and signaling pathways of phytohormones. Among the abiotic stresses, temperature stress (both high and low) limits the crop productivity worldwide. Crop production in low temperature is always a challenge, and sudden changes in temperatures to lower values have detrimental effect on crop productivity. For instance, in 2009, chilling temperature alone caused 158 billion yen damage in crop production in Japan. Additionally, early and late frost causes the damage of vegetable and fruits which is approximately 5–6 billion yen per year (Rahman 2013). With the predicted increase in population in coming years, and uncertain global weather pattern, it will be the future challenge of the plant scientists to develop crops resilient to temperature stress, including cold stress.

Low temperature affects the metabolic pathways and physiological development of plants. Cold stress is divided into nonfreezing or chilling stress (above 0°C and below 15°C) and freezing stress (below 0°C) (Shi and Yang 2014). However, in natural condition, temperature decreases gradually over time. Most of the temperate

M. A. Ashraf
United Graduate School of Agricultural Sciences, Iwate University, Morioka, Japan

A. Rahman (✉)
United Graduate School of Agricultural Sciences, Iwate University, Morioka, Japan

Department of Plant Bio Sciences, Faculty of Agriculture, Iwate University, Morioka, Japan

Agri-Innovation Center, Iwate University, Morioka, Japan
e-mail: abidur@iwate-u.ac.jp

S. H. Wani, V. Herath (eds.), *Cold Tolerance in Plants*,
https://doi.org/10.1007/978-3-030-01415-5_4

plants are capable of tolerating freezing stress after the exposure at nonfreezing or chilling condition for a certain amount of time. This is known as cold acclimation (CA) (Thomashow 1999). If plants are directly placed to freezing temperature without prior treatment or exposure at nonfreezing temperature, they are referred to as non-acclimated (NA) plants. During chilling stress and freezing stress conditions, plants go through diverse physiological and molecular changes. Chilling stress severely disrupts biochemical processes required for photosynthesis (Oliveira and Peñuelas 2005), reactive oxygen species (ROS) scavenging (Rizhsky et al. 2004), and hormonal homeostasis (Bielach et al. 2017; Shibasaki et al. 2009). These disruptions lead to cell death and subsequently crop damage (Hong et al. 2017). On the other hand, freezing stress induces ice crystal formation in the intracellular regions. Due to ice formation, cells perceive dehydration and osmotic stress, which eventually cause membrane and tissue injury (Uemura et al. 1995). In contrast, cold acclimated plants initiate the physiological and biochemical modifications. Additionally, cold-acclimated condition alters the expression patterns of cold-responsive genes, proteins, and metabolites (Miura and Furumoto 2013). As a result, cold-acclimated plants can survive during freezing temperature by accumulating osmolytes and antifreezing proteins (Yamada et al. 2002).

In this chapter, we will focus on hormonal regulation of cold stress and discuss about the recent advancement in this field and future strategies to develop plants that can combat low-temperature stress.

4.2 ABA (Abscisic Acid): The Major Stress Hormone

Abscisic acid (ABA) is an isoprenoid plant hormone produced in the plastid via 2-C methyl-D-erythritol-4-phosphate (MEP) pathway. It causes abscission of plant leaves, and its name reflects the function. It is a crucial phytohormone which plays important role in physiological and developmental stages including seed dormancy and development, stomatal opening, embryo morphogenesis, and responses under abiotic stress (Huang et al. 2017). Till date, it is the most studied phytohormone for its roles in facilitating plant adaptation against abiotic stresses. As a result, it has been termed as "stress hormone" (Wani et al. 2016).

4.2.1 Upregulation of ABA Biosynthesis in Cold Stress

Biosynthesis of ABA is augmented by cold stress which helps plants to withstand the adverse conditions (Gusta et al. 2005; Mega et al. 2015). In agreement with this observation, exogenous application of ABA stimulates freezing tolerance, and ABA biosynthesis mutants demonstrate cold sensitivity (Cuevas et al. 2008; Eremina et al. 2016). Increased biosynthesis of phytohormone reflects the upregulation of biosynthetic genes or downregulation of the genes in catabolic pathways. Under cold stress in *Arabidopsis* and rice, increased ABA level correlates with the induction of ABA biosynthetic pathway genes (Baron et al. 2012; Mega et al. 2015). Excessive ABA biosynthesis facilitates downstream events to bring changes at physiological level. For instance, exogenous ABA application decreases the electrolyte leakage (EL), malondialdehyde (MDA), and H_2O_2 content compared with plants without ABA treatment under cold treatment condition in Bermuda grass [*Cynodon dactylon* (L). Pers.] (Huang et al. 2017). EL is an indicator of membrane stability, MDA is a biomarker for oxidative damage, and H_2O_2 content indicates about oxidative stress.

Under cold stress, ABA content has been shown to be increased in cauline leaves and inflorescence meristem of *Arabidopsis thaliana* and shoot area of *Oryza sativa* L. (Baron et al. 2012; Mega et al. 2015). In *Arabidopsis*, ABA biosynthetic genes, catabolic genes are selectively overexpressed at 0°C in an organ-specific manner (Baron et al. 2012). For instance, *ABA1*, *ABA2*, *AAO3*, *NCED2*, *NCED5*, *NCED6*, *CYP707A1*, *CYP707A2*, *CYP707A3*, and *CYP707A4* in cauline leaves; *ABA1*, *ABA2*, *ABA4*, *AAO3*, *NCED3*, *CYO707A1*, *CYP707A2*, and *CYP707A4* in inflorescence meristem; and *ABA1*, *NCED2*, *NCED3*, *NCED5*, *NCED6*, *CYP707A2*, *CYP707A3*, and *CYP707A4* in developing silique are overexpressed during cold stress (Baron et al. 2012). Simultaneous induction of both biosynthetic and catabolic genes indicates the importance of intracellular ABA homeostasis. Phytohormone homeostasis involves transporters and enzymes responsible for the conjugation and hydrolysis reactions. ATP-binding cassette (ABC) protein ABCG25 and ABCG40 have been shown to export and import ABA, respectively (Kang et al. 2010; Kuromori et al. 2010). The most common conjugated form of ABA is ABA glucose ester (ABA-GE). UGT71B6 facilitates the conjugation to form ABA-GE. In contrast, AtBG1 acts on ABA-GE to produce free ABA through hydrolysis reaction (Baron et al. 2012).

Like *Arabidopsis*, cold stress-mediated induction of ABA content is also observed in rice after 24-h incubation at 15°C and 8°C. Even for the longer period of time up to 12 days at 15°C, the ABA content was high. Surprisingly, ABA content was not altered at 4°C after 24 h incubation and at 15°C for 15 days. This interesting finding was explained by the induction of expression of major catabolic gene, *OsABA0x1*, which increased the conjugated form of ABA, ABA-GE (Mega et al. 2015). Similarly, conjugating enzyme UGT71B6 (ABA-specific glucosyltransferase) was also found to be overexpressed in *Arabidopsis* under cold stress (Baron et al. 2012). Taken together, gene expression profile of biosynthetic, catabolic, transport, conjugation, and hydrolysis pathways suggests a net increase of ABA inside the cell under cold stress (Fig. 4.1).

4.2.2 Cold-Induced ABA Functions Through Both CBF-Dependent and CBF-Independent Pathways

Transcriptome studies identified set of genes that are altered during cold stress. These gene products have been shown to be involved in cold tolerance and response (Hannah et al. 2005; Lee et al. 2005; Shinozaki et al. 2003). The *Arabidopsis* genome consists of approximately 4% (Lee et al. 2005) to 20% (Hannah et al. 2005)

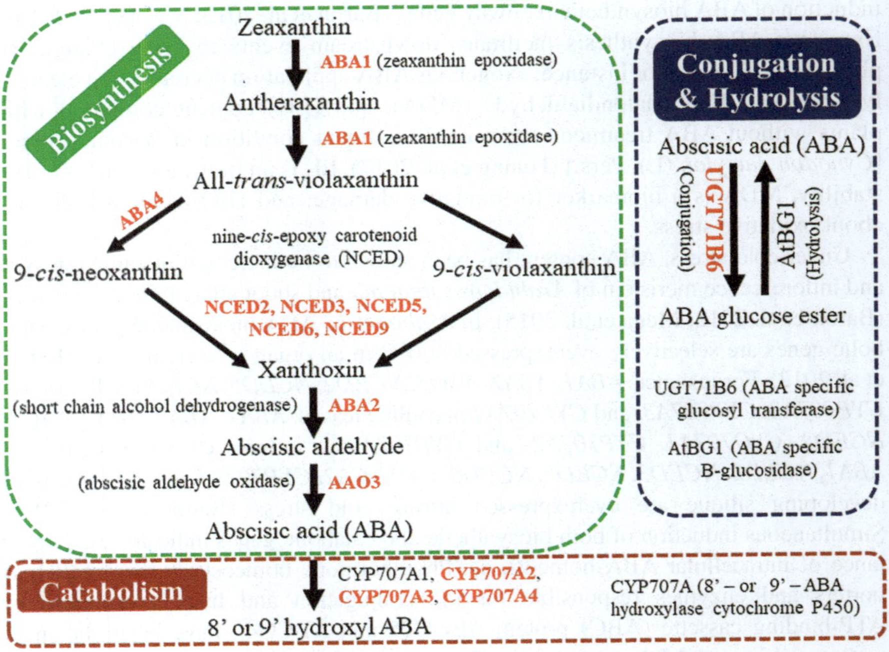

Fig. 4.1 Effect of cold stress on ABA biosynthesis, catabolism, conjugation, and hydrolysis pathway. Green, red and blue dotted boxes indicate biosynthesis, catabolism, conjugation, and hydrolysis pathway, respectively. Cold stress-induced genes are highlighted in red. The pathway is based on Baron et al. (2012). Organ-specific variation on gene expression has been mentioned in details in the text

cold-induced genes. Most of these genes contain C-repeat/dehydration-responsive element (*CRT/DRE*), *cis* regulatory motif, in promoter regions. The presence of these motifs facilitates the binding of drought-responsive element-binding (*DREB*)/C-Repeat Binding Factor (*CBF*) transcription factors which facilitate the response against cold stress in plant (Eremina et al. 2016). However, there are sets of genes which are altered by cold stress but not linked to CBF (Chinnusamy et al. 2007; Li et al. 2017; Xie et al. 2018). Hence, cold stress-mediated gene expression can be divided in two groups: CBF-dependent and CBF-independent.

As ABA does not alter CBF expression, it was hypothesized that ABA-mediated cold response follows CBF-independent pathway (Nakashima et al. 2014; Shinozaki and Yamaguchi-Shinozaki 2000). However, some recent results support the idea that ABA-mediated cold response may be mediated through CBF-dependent pathway. The transcription factor *MYB96* is induced by both ABA and cold (Lee and Seo 2015). Studies from loss of function and overexpression lines suggest that MYB96 contributes for cold tolerance (Lee and Seo 2015). Mechanistically, MYB96 interacts with HEPTAHELICAL PROTEIN (HHP). Interestingly, HHP1, HHP2, and HHP3 interact with ICE1, CAMTA, and ICE2, respectively (Chen et al. 2010; Lee and Seo 2015), supporting the notion that ABA-mediated cold stress resistance response is also regulated through CBF-dependent pathway (Fig. 4.3).

4.2.3 Epigenetic Control of ABA in Cold Stress

Gene expression is controlled by tight epigenetic regulation. DNA methylation and histone modifications such as acetylation, deacetylation, methylation, phosphorylation, and ubiquitination are major epigenetic regulations. For instance, histone acetylation activates transcription, and in opposite, histone deacetylation facilitates gene repression (Chen and Tian 2007). In addition, set of histone modification markers have been developed to understand whether the target gene is active or inactive. Histone H3 Lys9 (H3K9) acetylation, histone H3 Lys14 (H3K14) acetylation, and histone H3 Lys 4 (H3K4) trimethylation are known as markers for activated genes, while histone H3 Lys9 (H3K9) deacetylation, histone H3 Lys14 (H3K14) deacetylation, and histone H3 Lys9 (H3K9) demethylation are considered as markers for repressed genes (Chen and Tian 2007; Earley et al. 2006). HIGH EXPRESSION OF OSMOTICALLY RESPONSIVE GENE 15 (HOS15) encodes a WD-40 protein involved in histone deacetylation and had been reported as a repressor of cold tolerance-induced genes. Interestingly, both ABA and cold stress induce HOS15 expression, and loss-of-function mutant (*hos15–1*) is hypersensitive to freezing stress. Reduction in the transcripts of *CBF1*, *CBF2*, and *CBF3* is observed after cold treatment in *hos15* mutant (Zhu et al. 2008). HOS15 interacts with H4, and increased level of acetylated histone H4 is observed in *hos15–1* mutant (Zhu et al. 2008). This study provides the idea about the role of ABA-mediated histone deacetylation-dependent chromatin remodeling during cold stress. Furthermore, mutations in GCN5, encoding a histone acetyltransferase, which plays a role in determining the embryonic root-shoot axis, and ADA2, a transcriptional adaptor that interacts with histone acetyltransferase GCN5 homolog, alter the cold-regulated gene expression as well as cold tolerance (Vlachonasios et al. 2003).

4.3 Auxin: The Master Regulator of Plant Growth and Development

Auxin is the most studied and first discovered plant hormone (Enders and Strader 2015; Sauer et al. 2013; Tivendale and Cohen 2015). Biologically active major auxin is known as indole-3-acetic acid (IAA), and it controls almost all aspects of plant development from germination to senescence (Enders and Strader 2015; Rahman 2013; Sauer et al. 2013). The advent of genetic study based on mutants to decipher the function and use of the model plant *Arabidopsis thaliana* accelerated our progress to understand the auxin biosynthesis, signaling, and transport in the last few decades. As a fundamentally required hormone, unfortunately, our knowledge about auxin as stress hormone is limited. In recent years, several groups focused their research on understanding the roles of auxin in biotic and abiotic stresses.

Biosynthesis of IAA follows two pathways: tryptophan-dependent (Zhao 2010) and tryptophan-independent (Cohen et al. 2003; Strader and Bartel 2008; Zhao 2010). Tryptophan-dependent IAA biosynthesis is contributed by four pathways: IAOx (indole-3-acetaldoxime) pathway, IAM (indole-3-acetamide), IPA (indole-3-pyruvic acid) pathway, and YUC (YUCCA) pathway (Mashiguchi et al. 2011; Zhao 2010). Among these four pathways, IPA and YUC act as major IAA biosynthesis pathway (Zhao 2010). The major enzyme for IPA pathway is TAA1 (tryptophan aminotransferase of *Arabidopsis* 1), which converts tryptophan to IPA (Stepanova et al. 2008; Tao et al. 2008; Yamada et al. 2009). YUC genes work as flavin-containing monooxygenase for the conversion of IPA to IAA. Additionally, YUC genes are present in other species and play a crucial role in IAA-regulated developmental processes (Cheng et al. 2006; Cheng et al. 2007; Yamamoto et al. 2007; Zhao et al. 2001).

Auxin signaling starts with the perception of IAA by TIR1 (TRANSPORT INHIBITOR RESPONSE 1)/AFB (AUXIN SIGNALING F-BOX) receptors. Auxin acts as a molecular glue which brings together the transcriptional repressor Aux/IAAs and F-box proteins of the TIR1/AFB family. In usual condition, Aux/IAA is bound to ARF (auxin-responsive factor) and prevents the ARF-mediated expression of downstream genes. Upon auxin perception and ubiquitination of Aux/IAA proteins, ARFs induce the expression of auxin-regulated genes (Lavy and Estelle 2016; Leyser 2018).

4.3.1 Cellular Auxin Level Dictates the Survival of Cell and Cold Stress Tolerance

Intracellular concentration of biologically active indole-3-acetic acid (IAA) plays a major role in low-temperature stress. IAA concentration followed by cold stress is species- and organ-specific. For example, IAA concentration is increased in winter wheat, whereas it remained the same in spring wheat (Majláth et al. 2012). But study from the rice showed increased IAA content after cold stress treatment. The elevated amount of IAA is explained from molecular perspective due to the upregulation of auxin biosynthetic genes of the YUCCA family (*OsYUCCA2*, *OsYUCCA3*, *OsYUCCA6*, and *OsYUCCA7*) and downregulation of GH3 family (*OsGH3–1*, *OsGH3–2*, *OsGH3–5*, *OsGH3–6*, *OsGH3–7*, *OsGH3–9*, *OsGH3–11*, *OsGH3–13*), catalyzing IAA conjugation to amino acids (Du et al. 2013). Consistent with this evidence, overexpression of *OsGH3–2* shows IAA-deficient phenotypes such as dwarfism, smaller leaves, and fewer root hairs, but cold tolerance in rice. Taken together, the working model suggests that the unavailability of free IAA due to overexpression of *OsGH3–2* results in enhanced ROS scavenging events and induction of cold-responsive gene expression which provide membrane stability for cold tolerance (Du et al. 2012).

The importance of auxin level is also depicted as crucial point from another study during chilling stress. Recent work showed that chilling stress selectively facilitates the death of columella stem daughter cells (CSDCs) (Hong et al. 2017). It turned out that selective death of CSDCs helps to reestablish QC auxin maxima and improves the ability of the root to recover after chilling stress. Exogenous application of IAA reduces number of dead cells of CSDCs and improved root growth recovery. In contrast, biosynthetic inhibitor of auxin, yucasin, induces more CSDC death and makes root hypersensitive to recovery (Hong et al. 2017) (Fig. 4.2).

4.3.2 Cold Stress Alters Auxin Homeostasis by Inhibiting Auxin Transport

Intracellular auxin homeostasis is regulated by polar transport of auxin which depends on two streams of polar auxin transport, namely, rootward and shootward transports. Cold stress alters the intracellular auxin homeostasis by inhibiting shootward auxin transport, which was demonstrated by both direct transport assay and using the auxin- responsive marker *IAA2-GUS* (Shibasaki et al., 2009).

The shootward auxin transport is EIR1 (ETHYLENE INSENSITIVE ROOT 1)/ PIN2 (PIN-FORMED 2)-dependent. PIN2 is localized in the plasma membrane (PM) in a polar manner and continuously cycling back and forth between cytosol and PM (Adamowski and Friml 2015; Vieten et al. 2007). Inhibition of shootward auxin transport under cold stress was found to be associated with aberration of PIN2 activity. Interestingly, it was revealed that polar localization of PIN2 remains the same after cold stress, but the trafficking of PIN2 is affected (Shibasaki et al. 2009). Intracellular trafficking of PIN2 can be confirmed by applying the general protein trafficking inhibitor brefeldin A (BFA), which results in agglomeration of PIN2, commonly known as BFA bodies. Under cold stress, BFA-induced agglomeration is absent. Furthermore, during the recovery of root growth at optimal temperature after cold stress, the characteristic BFA bodies return which clearly suggests that cold stress specifically targets endosomal trafficking of PIN2 to reduce shootward auxin transport (Shibasaki et al. 2009) (Fig. 4.2). Consistent with this finding, it has been shown that expression of PINs in general is affected by cold stress (Hong et al. 2017). In addition, auxin maxima have been shown to play an important role in reestablishing the QC cells which facilitates the root growth recovery after cold stress (Hong et al. 2017). Collectively, these findings bring a new insight into the role of auxin in cold stress response.

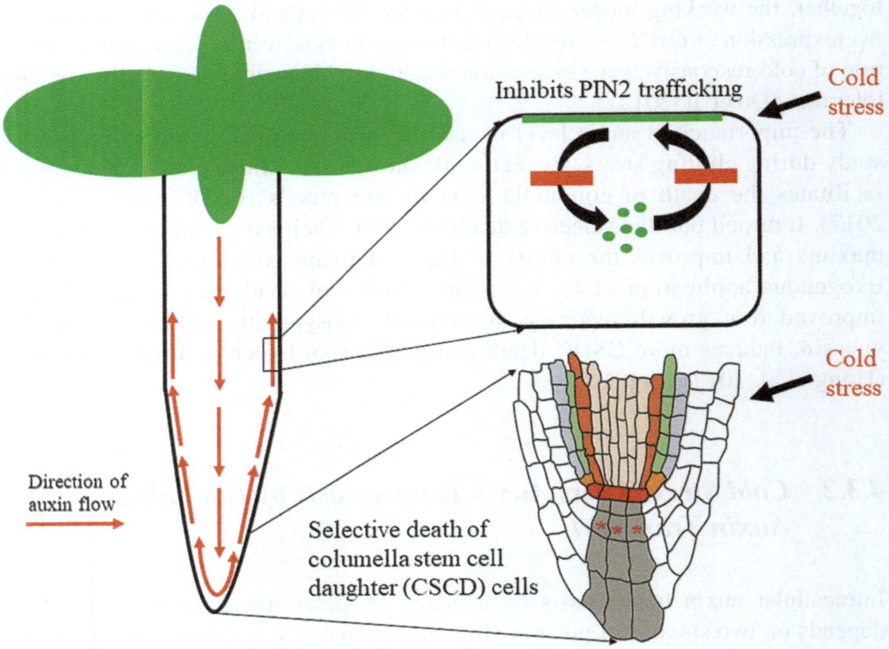

Fig. 4.2 Cold stress targets the endosomal trafficking of auxin efflux carrier, PIN2 (right side, upper panel). This figure is based on the finding of Shibasaki et al. (2009). Cold stress also induces selective death of columella stem cell daughter (CSCD) cells to reestablish the auxin maxima for survival (right side, lower panel, Hong et al. 2017)

4.4 Ethylene: The Gaseous Hormone for Ripening and Senescence

Ethylene, the simple two-carbon atom molecule, is the most characterized plant hormone for fruit ripening (Barry and Giovannoni 2007) and organ senescence (Burg 1968). Like auxin, the precursor of ethylene is amino acid methionine. The biosynthetic pathway starts with conversion of methionine to S-adenosyl-L-methionine (SAM) by SAM synthetase in an expenditure of ATP. SAM is converted to 5′-methylthioadenosine (MTA) and 1-amino cyclopropane-1-carboxylic acid (ACC) by the cleavage reaction facilitated by ACC synthase (ACS). MTA enters the Yang cycle, and ACC is converted to ethylene through an oxidation reaction by ACC oxidase (ACO) (Bakshi et al. 2015).

Ethylene is perceived by its receptors (ETR1, ETR2, ERS1, ERS2, and EIN4). Upon binding to the receptor, it activates the signaling cascade. CONSTITUTIVE TRIPLE RESPONSE1 (CTR1), negative regulator of the signaling pathway, controls the phosphorylation of NRAMP-like integral membrane protein ETHYLENE INSENSITIVE 2 (EIN2). C-terminal end of EIN2 (CEND EIN2) is cleaved and transduces signals from the cytoplasm to nucleus, which activates the downstream transcription factors (TFs) ETHYLENE INSENSITIVE 3 (EIN3) and EIN3-LIKE1

(EIL1) to regulate ethylene-mediated gene expression (Bakshi et al. 2015; Yang et al. 2015).

4.4.1 *Ethylene Regulates Cold Stress Response*

Ethylene content positively regulates freezing tolerance of non-acclimated plants in *Arabidopsis thaliana*. Exogenous application of ethylene precursor ACC results in higher survival rate under freezing stress (Catalá et al. 2014). The role of ethylene in freezing stress is further confirmed by the phenotype of ethylene-overproducing mutant, *eto1–3*, which shows enhanced freezing tolerance (Catalá and Salinas 2015). In contrast, the *acs* octuple mutant, containing mutations in 8 *ACS* genes (*ACS2*, *ACS4*, *ACS5*, *ACS6*, *ACS7*, *ACS8*, *ACS9*, and *ACS11*), which has ten times less ethylene content compared to wild type (Tsuchisaka et al. 2009), shows significantly lower survival rate upon freezing stress (Catalá et al. 2014). Consistently, octuple downregulates cold-induced gene (*CBF1*, *CBF2*, *RAP2.1*, *ERF4*, *ERF5*, *COR8.5*, *COR15A*, *LTI78*, *KIN1*, *COR47*, and *CHS*) expression (Catalá et al. 2014). Consistently, *eto1–3* upregulates cold-induced gene (*CBF1*, *CBF2*, *CBF3*) expression (Catalá and Salinas 2015). Altogether, it suggests that ethylene-mediated CBF-dependent pathway functions as positive regulator of freezing tolerance (Fig. 4.3).

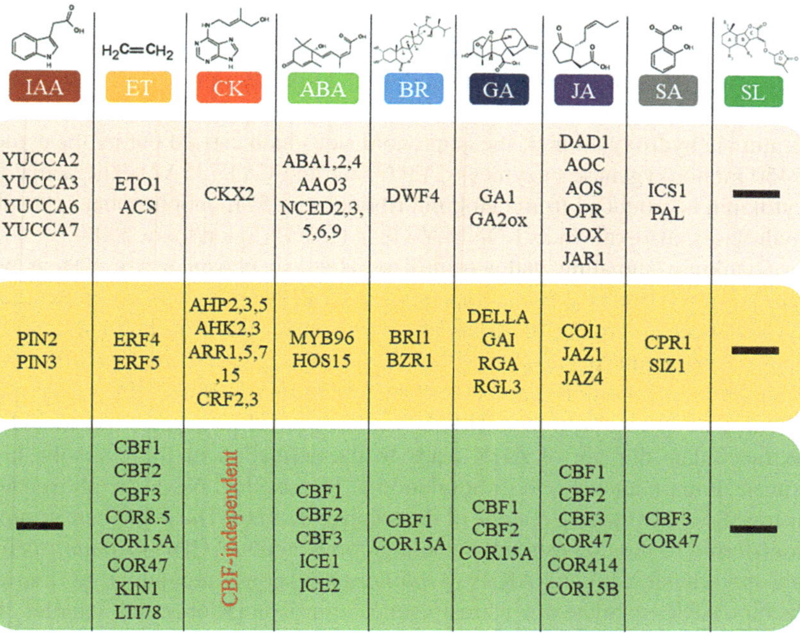

	IAA	ET	CK	ABA	BR	GA	JA	SA	SL
Biosynthesis/ catabolism	YUCCA2 YUCCA3 YUCCA6 YUCCA7	ETO1 ACS	CKX2	ABA1,2,4 AAO3 NCED2,3, 5,6,9	DWF4	GA1 GA2ox	DAD1 AOC AOS OPR LOX JAR1	ICS1 PAL	—
Signaling/ Transport	PIN2 PIN3	ERF4 ERF5	AHP2,3,5 AHK2,3 ARR1,5,7 ,15 CRF2,3	MYB96 HOS15	BRI1 BZR1	DELLA GAI RGA RGL3	COI1 JAZ1 JAZ4	CPR1 SIZ1	—
CBF/cold related genes	—	CBF1 CBF2 CBF3 COR8.5 COR15A COR47 KIN1 LTI78	CBF-independent	CBF1 CBF2 CBF3 ICE1 ICE2	CBF1 COR15A	CBF1 CBF2 COR15A	CBF1 CBF2 CBF3 COR47 COR414 COR15B	CBF3 COR47	—

Fig. 4.3 Present understanding about the involvement of different biosynthesis, transport, and signaling components of phytohormones and their involvements in regulating downstream signaling for known cold stress-regulated genes

Similar to *Arabidopsis*, an elevated level of ethylene has been observed in tomato (Ciardi et al. 1997), wheat (Kosová et al. 2012), alfalfa (Guo et al. 2014), and grapevine (Sun et al. 2016) under cold stress. Ethylene biosynthesis precursor ACC increases cold tolerance, and biosynthesis inhibitor AVG (aminoethoxyvinylglycine) reduces the cold tolerance in grapevine (Sun et al. 2016). Furthermore, it has been shown that low temperature induces the expression of *VaERF057*. In VaERF057-overexpressing transgenic line, *CBF1*, *CBF2*, and *CBF3* genes are upregulated which further suggests that ethylene-mediated CBF-dependent pathway is possibly a general mechanism by which plants respond to cold stress (Sun et al. 2016).

4.5 Cytokinin

Cytokinin was identified as a regulator of cell division from the initial screening (Miller et al. 1956; Miller et al. 1955). Chemically, cytokinin is the derivatives of adenine, where an isoprenoid side chain is attached at N^6 position. Cytokinin plays diverse roles in plant development such as embryo and gametophyte development; regulates pavement cell morphogenesis; inhibits lateral root formation; promotes nodulation, chloroplast development, and phloem development; regulates nutrient uptake; controls auxiliary bud release; inhibits senescence and cell proliferation in the root apical meristem; and induces cell division in the quiescent center (Kieber and Schaller 2018).

The first step of cytokinin biosynthesis starts with the addition of isopentenyl side chain from DMAPP (dimethylallyl diphosphate) to an adenosine moiety catalyzed by isopentenyl transferase (IPT). The resulting iP ribotides are converted to zeatin via hydroxylation of the isoprenoid side chain carried out by the cytochrome P450 monooxygenase enzymes (CYP735A1 and CYP735A2). The active form of cytokinin is produced from cytokinin ribotides by 5′-monophosphate phosphoribohydrolases, also known as LONELY GUY (LOG) (Sakakibara 2006).

Cytokinin signaling pathway utilizes His-Asp phosphorelay system which is similar to the bacterial two-component system (TCS). In this system, signal transduction occurs through the transfer of phosphate between the His residue of the sensor kinase and Asp residue of the receiver domain. In *Arabidopsis*, the cytokinin signaling starts with ER membrane-localized *Arabidopsis* histidine kinases (AHKs). Binding of cytokinin to the CHASE (cyclases/histidine kinases-associated sensing extracellular) domain of AHK leads to the activation of the cytosolic histidine-kinase domain and autophosphorylation at His residue. Next, the phosphate group is transferred to the Asp residue of the receiver domain. The phosphate group is later transferred to the downstream signaling components, AHPs (*Arabidopsis* histidine phosphotransferase) and ARRs (*Arabidopsis* response regulator). As transcription factors, ARRs regulate downstream gene expression (Kieber and Schaller 2018).

4.5.1 *Cytokinin Signaling Regulates Cold Stress in a CBF-Independent Pathway*

Cold stress induces expression of a subset of *ARRs* through AHK2 and AHK3 proteins without altering cytokinin levels. *ahk2 ahk3* and *ahk3 ahk4* mutants show enhanced freezing tolerance (Jeon et al. 2010). Consistently, overexpression of cold-induced *ARR7* demonstrates hypersensitivity, and *arr7* results in resistance to freezing stress (Jeon et al. 2010). This two-component system-mediated cold stress-responsive pathway is CBF-independent. But, *ahk2 ahk3* and *arr7* have hypersensitive response for ABA for seed germination. Altogether, it suggests a model of cold-responsive two-component signaling through the inhibition of ABA response (Jeon et al. 2010). *ARR1* and *AHP2*, *AHP3*, and *AHP5* play positive roles in cold-inducible expression of type A *ARRs*. For instance, cold stress induces the expression of type A ARR genes *ARR5*, *ARR6*, *ARR7*, and *ARR15*. But the induction of these cold-responsive type A ARR genes is absent in *arr1* (Jeon and Kim 2012), indicating that cold stress regulation is ARR1-dependent. This hypothesis is supported by the fact that *arr1* shows reduced survival rate and ARR1 overexpression results in tolerance to freezing stress (Jeon and Kim 2012). Additionally, it has been found that amino-terminal receiver domain of ARR1 is required for cold-responsive expression of type A *ARRs* (Jeon and Kim 2012).

Cytokinin response is also involved in lateral root (LR) initiation under cold stress. It has been revealed that *CRF2* (cytokinin response factor 2) and *CRF3* play a major role in lateral root initiation during cold stress. *crf2 crf3* produces a limited number of lateral roots under cold stress. Further it was demonstrated that CRF2 and CRF3 respond to cold through TCS-dependent and TCS-independent pathway, respectively (Jeon et al. 2016).

4.6 Brassinosteroid (BR)

Brassinosteroids (BRs) are important class of steroid plant hormones involved in cellular expansion and proliferation, vascular differentiation, male fertility, senescence, and leaf development (Clouse 2015; Fariduddin et al. 2014). The biosynthesis pathway of BRs starts with campesterol. The conversion of campesterol to BRs depends on the activities of two cytochrome P450s, DWF4 (DWARF 4) and CPD (CONSTITUTIVE PHOTOMORPHOGENESIS AND DWASRFISM) (Clouse 2015).

BR signaling starts with the binding of BR to the leucine-rich repeat receptor kinase (LRR-RK) BRI1 (BRASSINOSTEROID INSENSITIVE1) and activation of BRI1 kinase (Clouse et al. 1996; Li and Chory 1997). BRI1 activation facilitates the recruitment of co-receptor kinase BAK1 (BRI1-ASSOCIATED KINASE1) and dissociation of the BKI1 (BRI1 KINASE INHIBITOR1) (Li et al. 2002; Wang and Chory 2006). In BR signaling, BRI1 and BAK1 interaction works through "double-lock mechanism." When BRI1 is unoccupied by BR, extracellular and cytoplasmic domains of BRI1 prevent BAK1 interaction. In contrast, when the BRI1 is occupied by BR, phosphorylation of BKI1 causes it to dissociate and promotes the interaction of both extracellular and cytoplasmic domains between BRI1 and co-receptor BAK1 (Jaillais et al. 2011). Activated BRI1 kinase triggers a kinase cascade that activates transcription factor. Eventually BRI1 and BAK1 transphosphorylate the kinase domain of each other. After the activation of BRI1, it phosphorylates BSK1 (BRASSINOSTEROIDS-SIGNALING KINASE1) and CDG1 (CONSTITUTIVE DIFFERENTIAL GROWTH1). This phosphorylation initiates the BSK1 and CDG1 binding to BSU1 (BRI1-SUPPRESSOR1) and induces phosphorylation (Belkhadir et al. 2014). BSU1 acts as a phosphatase and dephosphorylates BIN2 (BRASSINOSTEROID INSENSITIVE2). At a low level of BR, active BIN2 phosphorylates transcription factors BZR1 (BRASSINAZOLE RESISTANT1) and BZR2/BES1 (BRI1-EMS-SUPPRESSOR1). In presence of BR, BIN2 is inactivated by BSU1 and degraded through proteasome. Subsequently, BZR1 and BZR2 are dephosphorylated by PP2A (PROTEIN PHOSPHATASE 2A) and regulate the expression of BR-regulated downstream genes (Belkhadir et al. 2014; Zhu et al. 2013).

BRs work as a positive regulator of cold tolerance in plants. Exogenous application of BRs helps to confer cold tolerance in maize (Singh et al. 2012) and cucumber (Jiang et al. 2013; Xia et al. 2009b). Overexpression of biosynthetic enzyme *DWF4* results in increased cold tolerance in *Arabidopsis* through inducing the expression of cold-responsive *COR15A* gene (Divi et al. 2016). Separate study demonstrated that the application of exogenous BRs enhances expression of *CBF1* (Kagale et al. 2007). Taken together, these results highlight the importance of BR in cold stress regulation and also establish that BR functions in cold stress through a CBF-dependent pathway (Fig. 4.3).

4.7 Gibberellic Acid (GA)

Gibberellic acid (GA) biosynthesis occurs in three stages. In the first stage, geranylgeranyl diphosphate is converted to ent-kaurene by ent-copalyl diphosphate synthase (CPS) and ent-kaurene synthase (KS) in chloroplast. In the next stage, ent-kaurene is converted to GA_{12} by ent-kaurene oxidase (KO) and ent-kaurenoic acid oxidase (KAO). These reactions take place at the plastid envelope and endoplasmic reticulum (ER). The final stage occurs in the cytosol where bioactive GAs, mostly GA_1 and GA_4, are produced by GA 20-oxidase (GA20ox) and GA 3-oxidase (GA3ox). Cellular homeostasis of GA is maintained by GA 2-oxidase. It adds a 2-OH group and prevents the binding of GA to its receptor (Yamaguchi 2008).

Cold stress affects both GA metabolic (Achard et al. 2008) and signaling pathways (Richter et al. 2013). Constitutive expression of *CBF1* shows freezing tolerances and accumulates DELLAs in *Arabidopsis*, which functions as a repressor of GA signaling (Achard et al. 2008). Accumulation of DELLA, which enhances the expression of *GA 2-oxidase* genes, results in reduction of bioactive GAs and GA content (Achard et al. 2008). Furthermore, overexpression of *CBF1/DREB1* shows chilling stress tolerance by restraining the bioactive GAs in tomato (Hsieh et al. 2002), *Capsella bursa-pastoris* (Zhou et al. 2014), and cotton (Shan et al. 2007).

GA biosynthetic mutant *ga1*, deficient in cyclase ent-kaurene synthetase A, shows freezing tolerance and also the induction of *CBF1*, *CBF2*, and *COR15A* genes (Fig. 4.3). These genes are induced by GA-mediated transcription factors GNC (GATA, NITRATE-INDUCIBLE, CARBON METABOLISM-INVOLVED) and GNL/CGA1 (GNC-LIKE/CYTOKININ RESPONSIVE GATA FACTOR1) (Richter et al. 2013; Richter et al. 2010). This model suggests the role of transcription factors GNC and GNL in freezing tolerance during low amount of bioactive GA content.

The importance of GA signaling in cold stress was demonstrated by using the GA-response mutant *gai*, double-DELLA (*gai-t6 rga-24*), and quadruple-DELLA (*gai-t6 rga-t2 rgl1–1 rgl2–1*) mutants (Achard et al. 2008). *Arabidopsis* contains five members of DELLA proteins: GIBBERELLIC ACID INSENSITIVE (GAI), REPRESSOR OF GAI (RGA), and RGA-LIKE 1, 2, and 3 (RGL1, RGL2, and RGL3). Genetic studies revealed that constitutive GA-signaling mutant *gai* shows freezing tolerance, and in contrast, DELLA knockout mutant *gai-t6 rga-24* shows sensitivity to freezing stress with and without cold acclimation (Achard et al. 2008). Interestingly, *CBF1* overexpression alters *RGL3* transcriptionally, suggesting a link between GA-signaling and CBF-dependent cold response pathway (Achard et al. 2008).

4.8 Jasmonic Acid (JA)

Jasmonic acid (JA) plays a vital role in the plant defense mechanism. In the pathogen-infected organ, JA works as a signal to initiate the response against pathogen (Wang and Wu 2013). The biosynthesis of jasmonic acid occurs in both plastid and peroxisome. In plastids, the biosynthesis of jasmonic acid starts with the peroxidation of α-linolenic acid by 13-lipoxygenase to form (13S)-hydroxyperoxyoctadecatrienoic acid (13-HPOT). Allene oxide synthase converts 13-HPOT into (13S)-12, 13-epoxyoctadecatrienoic acid (12, 13-EOT), and through the action of allene oxide cyclase, it eventually turns into cis-(+)-12-oxophytodienoic acid (OPDA). Later on, cis-(+)-OPDA is reduced by OPDA reductase and converted into 12-oxophytoenoic acid (OPC-8). In the following steps, through oxidation reaction, 12-carbon-containing jasmonic acid is produced. The free form of jasmonic acid (JA) and methylated form, methyl jasmonate (MeJA), are collectively known as jasmonates (JAs). JA is the most abundant in plant system, although it is not the active form. Isoleucine-conjugated JA (JA-Ile) is known as the most active form in *Arabidopsis* (Staswick and Tiryaki 2004). This conjugation reaction occurs in the cytosol by jasmonic acid-amido transferase 1 (JAR1). But, for the volatile defense signaling process, methylated form, MeJA is required, and the methylation takes place at cytosol by cytosolic methyl transferase (Wasternack and Hause 2013).

The JA signaling starts with the binding of JA-Ile to receptor molecule COI1 (CORONATINE INSENSITIVE 1). COI1 is an F-box protein, and upon binding with JA-Ile, it degrades repressor of JA, JAZ (JASMONATE ZIM DOMAIN) by ubiquitination (Wasternack and Hause 2013).

Endogenous JA level is increased due to cold treatment in wheat (Kosová et al. 2012), rice (Du et al. 2013), and *Arabidopsis* (Hu et al. 2013). Cold stress induces expression of *OsDAD1* (*DEFECTIVE ANTHER DEHISCENCE 1*), *OsAOC* (*ALLENE OXIDE CYCLASE*), *OsAOS1* (*ALLENE OXIDE SYNTHASE 1*), *OsAOS2*, *OsOPR1* (*12-OXOPHYTODIENOATE REDUCTASE 1*), *OsOPR7*, and *OsLOX2* (*LIPOOXYGENASE2*) from JA biosynthetic pathway and *OsCOI1a* (*CORONATINE INSENSITIVE 1*), *OsJAZ1* (*JASMONATE ZIM DOMAIN*), and *OsbHLH148* (*basic HELIX-LOOP-HELIX 148*) from JA signaling (Du et al. 2013). The observation in rice was also confirmed in *Arabidopsis*, where cold stress induces the expression of *AtLOX1*, *AtLOX2*, *AtLOX3*, *AtLOX4*, *AtAOS*, *AtAOC1*, *AtAOC2*, *AtAOC3*, *AtAOC4*, and *AtJAR1* (Hu et al. 2013). Exogenous application of JA helps to confer freezing tolerance and inhibition of endogenous jasmonate biosynthesis or signaling causes hypersensitive response to freezing stress (Hu et al. 2013).

Exogenous application of JA induces the freezing tolerance in a CBF-dependent pathway via JA signaling components. JA-signaling mutants (*jar1*, *coi1–1*, *coi1–2*) have reduced survival rate during freezing stress (Hu et al. 2013). JA induces expression of *CBF*s and CBF-regulated genes (*CBF1*, *CBF2*, *CBF3*, *COR47*, *COR414*, *COR15B*) after cold treatment (Hu et al. 2017; Sharma and Laxmi 2016) (Fig. 4.3). JA signaling components, JAZ1 and JAZ4, interact with ICE1 and ICE2. Overexpression of *JAZ1* and *JAZ4* represses the expression of *CBF*s and CBF-regulated genes on the downstream, and consistently, *JAZ1* and *JAZ4* overexpression results in reduced freezing tolerance before and after cold acclimation (Hu et al. 2013; Sharma and Laxmi 2016).

4.9 Salicylic Acid (SA)

2-Hydroxybenzoic acid or salicylic acid (SA) is a phenolic compound involved in plant growth, development, ripening, and pathogenesis-associated protein expression (Miura and Tada 2014). Although SA is studied extensively for defense response, it also helps to respond under abiotic stresses such as drought, chilling, heavy metal toxicity, heat, and osmotic stress (Rivas-San Vicente and Plasencia 2011). The SA biosynthesis follows two pathways: the isochorismate (IC) and the phenylalanine ammonia-lyase (PAL). Between them, IC pathway is the major pathway in *Arabidopsis thaliana* (Dempsey et al. 2011), *Nicotiana benthamiana* (Catinot et al. 2008), and tomato (Uppalapati et al. 2007).

Low temperature (8°C) induces the endogenous level of SA in cucumber (*Cucumis sativus* L.) through the PAL pathway based on differential gene expression of catalytic enzymes (Dong et al. 2014). Treatment with SA inhibitor results in severe chilling damage, which can be reverted back through the application of exogenous SA (Dong et al. 2014). Consistently, the application of SA alleviates chilling injury on peach fruit (Yang et al. 2012). In the model plant *Arabidopsis thaliana*, cold induces the accumulation of SA through the IC pathway. Low temperature induces the transcript of *ICS1*, and loss of function of mutant *ics1* is impaired to cold-induced SA biosynthesis (Kim et al. 2013).

In contrast to the above results, it has been shown that lower SA level facilitates the plant to withstand cold stress. Lower SA-containing transgenics *NahG* and *eds5* (*enhanced disease susceptibility 5*) grow faster at low temperature due to larger epidermal and mesophyll cell size (Scott et al. 2004; Xia et al. 2009a). In contrast, SA-over-accumulating line *cpr1* (*constitutive expression of PR genes*) shows reduced growth at low temperature (Scott et al. 2004). Additionally, other SA overproduction mutants *siz1–2* and *acd6* are hypersensitive to freezing stress with or without cold acclimation (Miura and Ohta 2010). Contrasting results are also available in the literature regarding the role of SA in cold stress and CBF pathway. It has been shown that in SA-deficient *NahG* line, the expression of *CBF1*, *CBF2*, and *CBF3* remains the same like wild type (Xia et al. 2009a). But a separate study showed that *CBF3* and *COR47* expressions are upregulated in SA-deficient *NahG* and downregulated in SA-overproducing *siz1–2* (Miura and Ohta 2010) (Fig. 4.3). Hence, the role of SA in cold stress may be condition- and species-specific.

4.10 Strigolactone (SL)

Strigolactone (SL) is a class of carotenoid-derived compounds. It was first isolated as seed germination stimulants in root parasitic plants of the family Orobanchaceae, including witchweeds (*Striga* spp.), broomrapes (*Orobanche* and *Phelipanche* spp.), and *Alectra* spp. (Ruyter-Spira et al. 2013; Xie et al. 2010). As a result, it was considered as harmful secondary metabolite for plant. But, later was found to act as necessary chemical signals for root colonization by symbiotic arbuscular mycorrhizal (AM) fungi (Akiyama et al. 2005). From that point, SL was considered as beneficial for plant. Additionally, recent report showed that SLs serve as important plant hormone to regulate shoot branching (Gomez-Roldan et al. 2008; Umehara et al. 2008).

SL is a relatively new plant hormone compared to others. Although its role has been reported for regulating root architecture, response under low phosphate conditions, and rhizosphere communication (Kapulnik and Koltai 2014), the involvement in cold stress response is yet to be elucidated.

4.11 Cross Talk Among Hormones to Regulate Low-Temperature Stress

A major issue with the hormonal regulation of plant development is its cross talk. Cross talk among hormones has made it complicated to interpret the effect of individual hormones. For instance, auxin and ethylene work synergistically to control root elongation and root hair formation but antagonistically for lateral root formation and hypocotyl elongation (Muday et al. 2012). The cross talk between cytokinin and auxin is mediated through SHY2. Cytokinin induces the expression of SHY2 through AHK3/ARR1 and eventually leads to the downregulation of *PIN1*, *PIN3*, and *PIN7*. On the other hand, auxin mediates SHY2 protein degradation through ubiquitination and releases PINs from negative regulation (Ioio et al. 2008).

Cytokinin induces the biosynthesis of ethylene by stabilizing the rate-limiting ACS (1-AMINOCYCLOPROPANE-1-CARBOXYLATE SYNTHASE) enzymes (Hansen et al. 2009; Vogel et al. 1998a; Vogel et al. 1998b). Cytokinin-induced inhibition of cell elongation is mediated by ethylene and AUX1; and inhibition of cell proliferation is mediated by ethylene and SHY2 (SHORT HYPOCOTYL 2)/IAA3 (Street et al. 2016). Like cytokinin, BR also stabilizes ACS from the ethylene biosynthesis pathway (Hansen et al. 2009) (Fig. 4.4).

Although it is not yet clear, possible hormonal cross talk exists in cold stress response too. For instance, cytokinin signaling components, ARR1 and ARR12, are involved in low- temperature-mediated root growth inhibition. At the same time, low temperature reduces the auxin biosynthesis and expression of auxin transporters (PIN1, PIN3, and PIN7) (Zhu et al. 2015). Opposite observations have been for auxin biosynthesis and transport reported in *arr1–3 arr12–1* background. *arr1–3 arr12–1* induces the expression of auxin biosynthesis genes *ASA1* and *YUC2* and transporter genes *PIN1*, *PIN3*, and *PIN7* (Zhu et al. 2015) (Fig. 4.4). These results suggest that cold stress uses cytokinin signaling component to inhibit auxin biosynthesis and transport for root growth inhibition.

From another study, it has been demonstrated that ethylene signaling negatively regulates freezing tolerance. Interestingly, transcription factor EIN3 negatively regulates the expression of cytokinin signaling component genes *ARR5*, *ARR7*, and *ARR15* (Shi et al. 2012). In fact, EIN3 binds directly to the promoter regions of cold-related genes (*CBF1*, *CBF2*, and *CBF3*) and also *ARR5*, *ARR7*, and *ARR15* (Fig. 4.4). Furthermore, exogenous application of ACC dramatically reduces the survival rate after freezing stress and downregulates transcriptional and protein levels of ARR5, ARR7, and ARR15. In contrast, increased freezing tolerance was observed using exogenous cytokinin and overexpressing *ARR5*, *ARR7*, and *ARR15* genes (Shi et al. 2012). Altogether these lines of evidence suggest a unified mechanism combining auxin, cytokinin, and ethylene interacting circuit as cold stress response regulators.

ARRs work as pivotal point to regulate freezing stress, and additional lines of evidence led to the hypothesis that these signaling components not only act on the convergence of auxin, cytokinin, and ethylene but are also connected with master

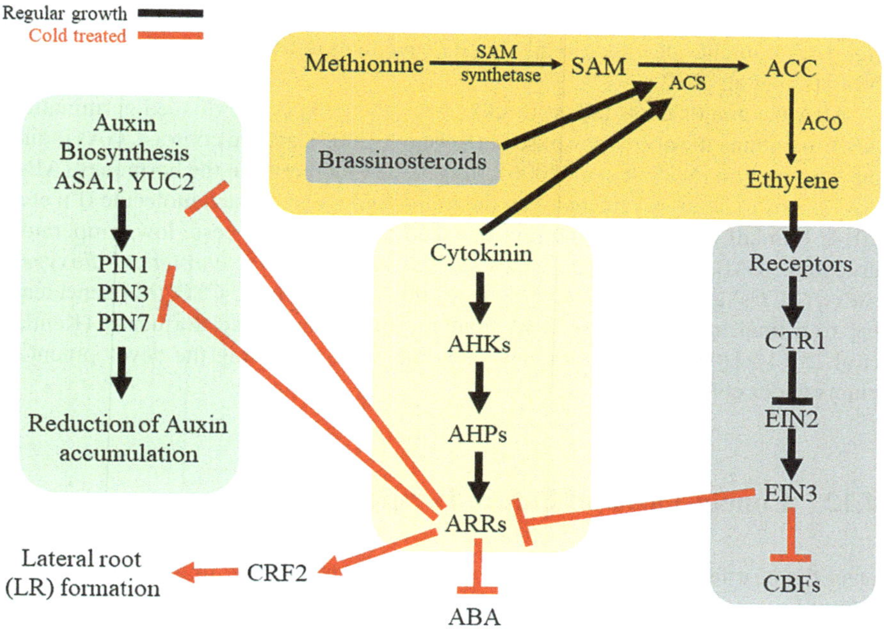

Fig. 4.4 Hormonal cross talk among auxin, ethylene, cytokinin, ABA, and BRs under cold stress. Cytokinin and BRs stabilize the rate-limiting enzyme, ACS, of ethylene biosynthesis during regular growth condition. EIN3 negatively regulates CBFs and group of ARRs under cold stress. A group of ARRs have multiple functions: inhibit ABA response, regulate CRF2-mediated lateral root formation, downregulate auxin transporters, and reduce auxin biosynthesis under low-temperature stress. Black and red arrows correspond to regular growth condition and cold treatment, respectively. Specific signaling components and enzymes are mentioned in detail in the text

stress hormone ABA. ABA and cytokinin signaling have inverse relationship during cold stress. Cold stress increases ABA signaling and decreases cytokinin signaling. High levels of *Oryza sativa 9-cis-epoxycarotenoid dioxygenase* transcripts correlate with ABA accumulation, and low levels of *Cytochrome P450 (CYP) 735A* transcripts correlate with decreased levels of a CK precursor in rice. This reduced expression of *CYP735A*s occurs in rice but not in *Arabidopsis* (Maruyama et al. 2014). These results highlight the differences of cytokinin signaling between monocot and dicot under cold stress. Cytokinin-signaling mutants *ahk2 ahk3* and *arr7* have hypersensitive response for ABA for seed germination, which suggests a possible model of cold-responsive cytokinin signaling through the inhibition of ABA response (Jeon et al. 2010) (Fig. 4.4).

The potential cross talk between IAA and ABA in cold tolerance was studied in *OsGH3–2* overexpressing rice lines (an enzyme that catalyzes IAA conjugation into amino acids) where a reduction in free IAA content caused a concomitant increase in ABA levels and in cold tolerance (Du et al. 2012). Overexpression of *OsGH3–2* in rice caused significant morphological aberrations related to IAA deficiency, such as dwarfism, smaller leaves, and fewer crown roots and root hairs. The overexpression line showed increased cold tolerance, which was due to combined effects of reduced free IAA content, alleviated oxidative damage, and decreased membrane penetrability (Du et al. 2012).

ABA has major cross talk with GAs from the perspective of seed germination. ABA maintains the dormancy of seed (Kermode 2005), and, in contrast, GAs induce the germination (Nadjafi et al. 2006). In fact, the ratio between the hormones (ABA/GA) seems to be more relevant than the individual effect of each molecule (Fu et al. 2014; Kendall et al. 2011). During the seed maturation process, low temperature upregulates ABA biosynthetic genes such as *nine-cis-epoxy carotenoid dioxygenases* (*NECD*s) and strongly downregulates the ABA-catabolic *CYP707A* gene, leading to a final increase in the ABA content, maintaining seed dormancy (Kendall et al. 2011). However, it is not clear how they interact during the development of plants under cold stress.

4.12 Conclusions and Future Perspectives

Over the past few decades, major progresses have been made in understanding the role of hormonal interplay for low-temperature response (Fig. 4.3). Even after that, our understanding still lacks comprehensive mechanistic explanation for the role of hormones in cold stress response. This is further complicated by the cross talk between the hormones. The understanding of hormonal regulation of cold stress response indicates two distinct pathways: one is highly explored CBF-dependent pathway and the other one is less studied CBF-independent pathway. Understanding the CBF-independent pathway in detail will provide a new insight into the cold stress response mechanism. One potential CBF-independent pathway is protein trafficking. The hormonal regulation of plant is subjected to intracellular protein trafficking (Adamowski and Friml 2015; Jaillais and Vert 2016; Löfke et al. 2013; Salanenka et al. 2018), and cold stress specifically alters the trafficking of a subset of proteins which affects the hormonal homeostasis inside the cell (Shibasaki et al. 2009). Elucidating these specific changes and the target proteins will further enhance our understanding of the cold stress response mechanism, and combining the knowledge of CBF-dependent and CBF-independent pathways will help us to make plants resilient to cold stress in the future.

In plants, multiple processes such as genome imprinting, stress responses, and cellular differentiation are mediated through epigenetic regulation. Epigenetics control gene expression and genome integrity where DNA sequence remains same. Few common epigenetic regulations are DNA methylation, histone methylation, histone acetylation etc. (Takatsuka and Umeda 2015). These epigenetic modifications regulate hormonal signaling and vice versa (Yamamuro et al. 2016). Till date, ethylene is the most studied phytohormone for the epigenetic regulations (Wang et al. 2017; Zhang et al. 2018; Zhang et al. 2017). Apart from that, auxin, cytokinin, ABA, BR, GA, and JA responses are also directly or indirectly regulated by epigenetic-related factors (Yamamuro et al. 2016). On the other hand, the epigenetic regulation of cold stress response has started to emerge (Banerjee et al. 2017). How the epigenetic control or "memory response" is integrated with hormonal signaling and dictates developmental cue of plant under low- temperature stress will be an intriguing area of future research.

Acknowledgment Research in Abidur Rahman's lab is funded by several grants by JSPS Kakenhi and Iwate University. Arif Ashraf was supported by MEXT fellowship.

References

Achard P, Gong F, Cheminant S, Alioua M, Hedden P, Genschik P (2008) The cold-inducible CBF1 factor–dependent signaling pathway modulates the accumulation of the growth-repressing DELLA proteins via its effect on gibberellin metabolism. Plant Cell 20:2117–2129

Adamowski M, Friml J (2015) PIN-dependent auxin transport: action, regulation, and evolution. Plant Cell 27:20–32

Akiyama K, Matsuzaki K, Hayashi H (2005) Plant sesquiterpenes induce hyphal branching in arbuscular mycorrhizal fungi. Nature 435:824

Bakshi A, Shemansky JM, Chang C, Binder BM (2015) History of research on the plant hormone ethylene. J Plant Growth Regul 34:809–827

Banerjee A, Wani SH, Roychoudhury A (2017) Epigenetic control of plant cold responses. Front Plant Sci 8:1643

Baron KN, Schroeder DF, Stasolla C (2012) Transcriptional response of abscisic acid (ABA) metabolism and transport to cold and heat stress applied at the reproductive stage of development in Arabidopsis thaliana. Plant Sci 188:48–59

Barry CS, Giovannoni JJ (2007) Ethylene and fruit ripening. J Plant Growth Regul 26:143

Belkhadir Y, Yang L, Hetzel J, Dangl JL, Chory J (2014) The growth–defense pivot: crisis management in plants mediated by LRR-RK surface receptors. Trends Biochem Sci 39:447–456

Bielach A, Hrtyan M, Tognetti VB (2017) Plants under stress: involvement of auxin and cytokinin. Int J Mol Sci 18:1427

Burg SP (1968) Ethylene, plant senescence and abscission. Plant Physiol 43:1503

Catalá R, López-Cobollo R, Castellano MM, Angosto T, Alonso JM, Ecker JR, Salinas J (2014) The Arabidopsis 14-3-3 protein RARE COLD INDUCIBLE 1A links low-temperature response and ethylene biosynthesis to regulate freezing tolerance and cold acclimation. Plant Cell 26:3326–3342

Catalá R, Salinas J (2015) The Arabidopsis ethylene overproducer mutant eto1-3 displays enhanced freezing tolerance. Plant Signal Behav 10:e989768

Catinot J, Buchala A, Abou-Mansour E, Métraux J-P (2008) Salicylic acid production in response to biotic and abiotic stress depends on isochorismate in Nicotiana benthamiana. FEBS Lett 582:473–478

Chen C-C, Liang C-S, Kao A-L, Yang C-C (2010) HHP1, a novel signalling component in the cross-talk between the cold and osmotic signalling pathways in Arabidopsis. J Exp Bot 61:3305–3320

Chen ZJ, Tian L (2007) Roles of dynamic and reversible histone acetylation in plant development and polyploidy. Biochimica et Biophysica Acta (BBA)-Gene Structure and Expression 1769:295–307

Cheng Y, Dai X, Zhao Y (2006) Auxin biosynthesis by the YUCCA flavin monooxygenases controls the formation of floral organs and vascular tissues in Arabidopsis. Genes Dev 20:1790–1799

Cheng Y, Dai X, Zhao Y (2007) Auxin synthesized by the YUCCA flavin monooxygenases is essential for embryogenesis and leaf formation in Arabidopsis. Plant Cell 19:2430–2439
Chinnusamy V, Zhu J, Zhu J-K (2007) Cold stress regulation of gene expression in plants. Trends Plant Sci 12:444–451
Ciardi JA, Deikman J, Orzolek MD (1997) Increased ethylene synthesis enhances chilling tolerance in tomato. Physiol Plant 101:333–340
Clouse SD (2015) A history of brassinosteroid research from 1970 through 2005: thirty-five years of phytochemistry, physiology, genes, and mutants. J Plant Growth Regul 34:828–844
Clouse SD, Langford M, McMorris TC (1996) A brassinosteroid-insensitive mutant in Arabidopsis thaliana exhibits multiple defects in growth and development. Plant Physiol 111:671–678
Cohen JD, Slovin JP, Hendrickson AM (2003) Two genetically discrete pathways convert tryptophan to auxin: more redundancy in auxin biosynthesis. Trends Plant Sci 8:197–199
Cuevas JC et al (2008) Putrescine is involved in Arabidopsis freezing tolerance and cold acclimation by regulating abscisic acid levels in response to low temperature. Plant Physiol 148:1094–1105
Dempsey DMA, Vlot AC, Wildermuth MC, Klessig DF (2011) Salicylic acid biosynthesis and metabolism. The Arabidopsis Book 9:e0156
Divi UK, Rahman T, Krishna P (2016) Gene expression and functional analyses in brassinosteroid-mediated stress tolerance. Plant Biotechnol J 14:419–432
Dong C-J, Li L, Shang Q-M, Liu X-Y, Zhang Z-G (2014) Endogenous salicylic acid accumulation is required for chilling tolerance in cucumber (Cucumis sativus L.) seedlings. Planta 240:687–700
Du H, Liu H, Xiong L (2013) Endogenous auxin and jasmonic acid levels are differentially modulated by abiotic stresses in rice. Front Plant Sci 4:397
Du H, Wu N, Fu J, Wang S, Li X, Xiao J, Xiong L (2012) A GH3 family member, OsGH3-2, modulates auxin and abscisic acid levels and differentially affects drought and cold tolerance in rice. J Exp Bot 63:6467–6480
Earley K et al (2006) Erasure of histone acetylation by Arabidopsis HDA6 mediates large-scale gene silencing in nucleolar dominance. Genes Dev 20:1283–1293
Enders TA, Strader LC (2015) Auxin activity: past, present, and future. Am J Bot 102:180–196
Eremina M, Rozhon W, Poppenberger B (2016) Hormonal control of cold stress responses in plants. Cell Mol Life Sci 73:797–810
Fariduddin Q, Yusuf M, Ahmad I, Ahmad A (2014) Brassinosteroids and their role in response of plants to abiotic stresses. Biol Plant 58:9–17
Fu X, Liu H, Xu J, Tang J, Shang X (2014) Primary metabolite mobilization and hormonal regulation during seed dormancy release in Cornus japonica var. chinensis. Scand J For Res 29:542–551
Gomez-Roldan V et al (2008) Strigolactone inhibition of shoot branching. Nature 455:189
Guo Z, Tan J, Zhuo C, Wang C, Xiang B, Wang Z (2014) Abscisic acid, H2O2 and nitric oxide interactions mediated cold-induced S-adenosylmethionine synthetase in Medicago sativa subsp. falcata that confers cold tolerance through up-regulating polyamine oxidation. Plant Biotechnol J 12:601–612
Gusta L, Trischuk R, Weiser C (2005) Plant cold acclimation: the role of abscisic acid. J Plant Growth Regul 24:308–318
Hannah MA, Heyer AG, Hincha DK (2005) A global survey of gene regulation during cold acclimation in Arabidopsis thaliana. PLoS Genet 1:e26
Hansen M, Chae HS, Kieber JJ (2009) Regulation of ACS protein stability by cytokinin and brassinosteroid. Plant J 57:606–614
Hong JH et al (2017) A sacrifice-for-survival mechanism protects root stem cell niche from chilling stress. Cell 170:102–113 e114.
Hsieh T-H, Lee J-T, Yang P-T, Chiu L-H, Y-y C, Wang Y-C, Chan M-T (2002) Heterology expression of the ArabidopsisC-repeat/dehydration response element binding factor 1 gene confers elevated tolerance to chilling and oxidative stresses in transgenic tomato. Plant Physiol 129:1086–1094

Hu Y, Jiang L, Wang F, Yu D (2013) Jasmonate regulates the inducer of CBF expression–c-repeat binding factor/DRE binding factor1 cascade and freezing tolerance in Arabidopsis. Plant Cell 25:2907–2924

Hu Y, Jiang Y, Han X, Wang H, Pan J, Yu D (2017) Jasmonate regulates leaf senescence and tolerance to cold stress: crosstalk with other phytohormones. J Exp Bot 68:1361–1369

Huang X, Shi H, Hu Z, Liu A, Amombo E, Chen L, Fu J (2017) ABA is involved in regulation of cold stress response in Bermudagrass. Front Plant Sci 8:1613

Ioio RD et al (2008) A genetic framework for the control of cell division and differentiation in the root meristem. Science 322:1380–1384

Jaillais Y, Belkhadir Y, Balsemão-Pires E, Dangl JL, Chory J (2011) Extracellular leucine-rich repeats as a platform for receptor/coreceptor complex formation. Proc Natl Acad Sci 108:8503–8507

Jaillais Y, Vert G (2016) Brassinosteroid signaling and BRI1 dynamics went underground. Curr Opin Plant Biol 33:92–100

Jeon J, Cho C, Lee MR, Van Binh N, Kim J (2016) CYTOKININ RESPONSE FACTOR2 (CRF2) and CRF3 regulate lateral root development and response to cold stress in Arabidopsis. Plant Cell. https://doi.org/10.1105/tpc.15.00909

Jeon J, Kim J (2012) Arabidopsis response regulator 1 (ARR1) and Arabidopsis histidine phosphotransfer protein 2 (AHP2), AHP3, and AHP5 function in cold signaling. Plant Physiol. https://doi.org/10.1104/pp.112.207621

Jeon J et al (2010) A subset of cytokinin two-component signaling system plays a role in cold temperature stress response in Arabidopsis. J Biol Chem 285:23371–23386

Jiang YP et al (2013) Brassinosteroids accelerate recovery of photosynthetic apparatus from cold stress by balancing the electron partitioning, carboxylation and redox homeostasis in cucumber. Physiol Plant 148:133–145

Kagale S, Divi UK, Krochko JE, Keller WA, Krishna P (2007) Brassinosteroid confers tolerance in Arabidopsis thaliana and Brassica napus to a range of abiotic stresses. Planta 225:353–364

Kang J, Hwang J-U, Lee M, Kim Y-Y, Assmann SM, Martinoia E, Lee Y (2010) PDR-type ABC transporter mediates cellular uptake of the phytohormone abscisic acid. Proc Natl Acad Sci 107:2355–2360

Kapulnik Y, Koltai H (2014) Strigolactone involvement in root development, response to abiotic stress, and interactions with the biotic soil environment. Plant Physiol 166:560–569

Kendall SL, Hellwege A, Marriot P, Whalley C, Graham IA, Penfield S (2011) Induction of dormancy in Arabidopsis summer annuals requires parallel regulation of DOG1 and hormone metabolism by low temperature and CBF transcription factors. Plant Cell 23:2568–2580

Kermode AR (2005) Role of abscisic acid in seed dormancy. J Plant Growth Regul 24:319–344

Kieber JJ, Schaller GE (2018) Cytokinin signaling in plant development. Development 145:dev149344

Kim Y, Park S, Gilmour SJ, Thomashow MF (2013) Roles of CAMTA transcription factors and salicylic acid in configuring the low-temperature transcriptome and freezing tolerance of Arabidopsis. Plant J 75:364–376

Kosová K et al (2012) Complex phytohormone responses during the cold acclimation of two wheat cultivars differing in cold tolerance, winter Samanta and spring Sandra. J Plant Physiol 169:567–576

Kuromori T et al (2010) ABC transporter AtABCG25 is involved in abscisic acid transport and responses. Proc Natl Acad Sci 107:2361–2366

Lavy M, Estelle M (2016) Mechanisms of auxin signaling. Development 143:3226–3229

Lee B-h, Henderson DA, Zhu J-K (2005) The Arabidopsis cold-responsive transcriptome and its regulation by ICE1. Plant Cell 17:3155–3175

Lee HG, Seo PJ (2015) The MYB96–HHP module integrates cold and abscisic acid signaling to activate the CBF–COR pathway in Arabidopsis. Plant J 82:962–977

Leyser O (2018) Auxin signaling. Plant Physiol 176:465–479

Li H, Ye K, Shi Y, Cheng J, Zhang X, Yang S (2017) BZR1 positively regulates freezing tolerance via CBF-dependent and CBF-independent pathways in Arabidopsis. Mol Plant 10:545–559

Li J, Chory J (1997) A putative leucine-rich repeat receptor kinase involved in brassinosteroid signal transduction. Cell 90:929–938

Li J, Wen J, Lease KA, Doke JT, Tax FE, Walker JC (2002) BAK1, an Arabidopsis LRR receptor-like protein kinase, interacts with BRI1 and modulates brassinosteroid signaling. Cell 110:213–222

Löfke C, Zwiewka M, Heilmann I, Van Montagu MC, Teichmann T, Friml J (2013) Asymmetric gibberellin signaling regulates vacuolar trafficking of PIN auxin transporters during root gravitropism. Proc Natl Acad Sci 110:3627–3632

Majláth I et al (2012) Effect of light on the gene expression and hormonal status of winter and spring wheat plants during cold hardening. Physiol Plant 145:296–314

Maruyama K et al (2014) Integrated analysis of the effects of cold and dehydration on rice metabolites, phytohormones, and gene transcripts. Plant Physiol 164:1759–1771

Mashiguchi K et al (2011) The main auxin biosynthesis pathway in Arabidopsis. Proc Natl Acad Sci 108:18512–18517

Mega R et al (2015) Sustained low abscisic acid levels increase seedling vigor under cold stress in rice (Oryza sativa L.). Sci Rep 5:13819

Miller CO, Skoog F, Okumura F, Von Saltza M, Strong F (1956) Isolation, structure and synthesis of kinetin, a substance promoting cell division1, 2. J Am Chem Soc 78:1375–1380

Miller CO, Skoog F, Von Saltza MH, Strong F (1955) Kinetin, a cell division factor from deoxyribonucleic acid1. J Am Chem Soc 77:1392–1392

Miura K, Furumoto T (2013) Cold signaling and cold response in plants. Int J Mol Sci 14:5312–5337

Miura K, Ohta M (2010) SIZ1, a small ubiquitin-related modifier ligase, controls cold signaling through regulation of salicylic acid accumulation. J Plant Physiol 167:555–560

Miura K, Tada Y (2014) Regulation of water, salinity, and cold stress responses by salicylic acid. Front Plant Sci 5:4

Muday GK, Rahman A, Binder BM (2012) Auxin and ethylene: collaborators or competitors? Trends Plant Sci 17:181–195

Nadjafi F, Bannayan M, Tabrizi L, Rastgoo M (2006) Seed germination and dormancy breaking techniques for Ferula gummosa and Teucrium polium. J Arid Environ 64:542–547

Nakashima K, Yamaguchi-Shinozaki K, Shinozaki K (2014) The transcriptional regulatory network in the drought response and its crosstalk in abiotic stress responses including drought, cold, and heat. Front Plant Sci 5

Oliveira G, Peñuelas J (2005) Effects of winter cold stress on photosynthesis and photochemical efficiency of PSII of the Mediterranean Cistus albidus L. and Quercus ilex L. Plant Ecol 175:179–191

Rahman A (2013) Auxin: a regulator of cold stress response. Physiol Plant 147:28–35

Richter R, Bastakis E, Schwechheimer C (2013) Cross-repressive interactions between SOC1 and the GATAs GNC and GNL/CGA1 in the control of greening, cold tolerance, and flowering time in Arabidopsis. Plant Physiol 162:1992–2004

Richter R, Behringer C, Müller IK, Schwechheimer C (2010) The GATA-type transcription factors GNC and GNL/CGA1 repress gibberellin signaling downstream from DELLA proteins and PHYTOCHROME-INTERACTING FACTORS. Genes Dev 24:2093–2104

Rivas-San Vicente M, Plasencia J (2011) Salicylic acid beyond defence: its role in plant growth and development. J Exp Bot 62:3321–3338

Rizhsky L, Davletova S, Liang H, Mittler R (2004) The zinc finger protein Zat12 is required for cytosolic ascorbate peroxidase 1 expression during oxidative stress in Arabidopsis. J Biol Chem 279:11736–11743

Ruyter-Spira C, Al-Babili S, Van Der Krol S, Bouwmeester H (2013) The biology of strigolactones. Trends Plant Sci 18:72–83

Sakakibara H (2006) Cytokinins: activity, biosynthesis, and translocation. Annu Rev Plant Biol 57:431–449

Salanenka Y, Verstraeten I, Löfke C, Tabata K, Naramoto S, Glanc M, Friml J (2018) Gibberellin DELLA signaling targets the retromer complex to redirect protein trafficking to the plasma membrane. Proc Natl Acad Sci:201721760. https://doi.org/10.1073/pnas.1721760115

Sauer M, Robert S, Kleine-Vehn J (2013) Auxin: simply complicated. J Exp Bot 64:2565–2577
Scott IM, Clarke SM, Wood JE, Mur LA (2004) Salicylate accumulation inhibits growth at chilling temperature in Arabidopsis. Plant Physiol 135:1040–1049
Shan DP et al (2007) Cotton GhDREB1 increases plant tolerance to low temperature and is negatively regulated by gibberellic acid. New Phytol 176:70–81
Sharma M, Laxmi A (2016) Jasmonates: emerging players in controlling temperature stress tolerance. Front Plant Sci 6:1129
Shi Y, Tian S, Hou L, Huang X, Zhang X, Guo H, Yang S (2012) Ethylene signaling negatively regulates freezing tolerance by repressing expression of CBF and type-a ARR genes in Arabidopsis. Plant Cell 24:2578–2595
Shi Y, Yang S (2014) ABA regulation of the cold stress response in plants. In: Abscisic acid: metabolism, transport and signaling. Springer, Dordrecht, pp 337–363
Shibasaki K, Uemura M, Tsurumi S, Rahman A (2009) Auxin response in Arabidopsis under cold stress: underlying molecular mechanisms. Plant Cell 21:3823–3838
Shinozaki K, Yamaguchi-Shinozaki K (2000) Molecular responses to dehydration and low temperature: differences and cross-talk between two stress signaling pathways. Curr Opin Plant Biol 3:217–223
Shinozaki K, Yamaguchi-Shinozaki K, Seki M (2003) Regulatory network of gene expression in the drought and cold stress responses. Curr Opin Plant Biol 6:410–417
Singh I, Kumar U, Singh S, Gupta C, Singh M, Kushwaha S (2012) Physiological and biochemical effect of 24-epibrassinoslide on cold tolerance in maize seedlings. Physiol Mol Biol Plants 18:229–236
Staswick PE, Tiryaki I (2004) The oxylipin signal jasmonic acid is activated by an enzyme that conjugates it to isoleucine in Arabidopsis. Plant Cell 16:2117–2127
Stepanova AN et al (2008) TAA1-mediated auxin biosynthesis is essential for hormone crosstalk and plant development. Cell 133:177–191
Strader LC, Bartel B (2008) A new path to auxin. Nat Chem Biol 4:337
Street IH et al (2016) Cytokinin acts through the auxin influx carrier AUX1 to regulate cell elongation in the root. Development 143:3982–3993
Sun X et al (2016) Ethylene positively regulates cold tolerance in grapevine by modulating the expression of ETHYLENE RESPONSE FACTOR 057. Sci Rep 6:24066
Takatsuka H, Umeda M (2015) Epigenetic control of cell division and cell differentiation in the root apex. Front Plant Sci 6:1178
Tao Y et al (2008) Rapid synthesis of auxin via a new tryptophan-dependent pathway is required for shade avoidance in plants. Cell 133:164–176
Thomashow MF (1999) Plant cold acclimation: freezing tolerance genes and regulatory mechanisms. Annu Rev Plant Biol 50:571–599
Tivendale ND, Cohen JD (2015) Analytical history of auxin. J Plant Growth Regul 34:708–722
Tsuchisaka A et al (2009) A combinatorial interplay among the 1-aminocyclopropane-1-carboxylate isoforms regulates ethylene biosynthesis in Arabidopsis thaliana. Genetics 183:979–1003
Uemura M, Joseph RA, Steponkus PL (1995) Cold acclimation of Arabidopsis thaliana (effect on plasma membrane lipid composition and freeze-induced lesions). Plant Physiol 109:15–30
Umehara M et al (2008) Inhibition of shoot branching by new terpenoid plant hormones. Nature 455:195
Uppalapati SR, Ishiga Y, Wangdi T, Kunkel BN, Anand A, Mysore KS, Bender CL (2007) The phytotoxin coronatine contributes to pathogen fitness and is required for suppression of salicylic acid accumulation in tomato inoculated with Pseudomonas syringae pv. Tomato DC3000. Mol Plant-Microbe Interact 20:955–965
Vieten A, Sauer M, Brewer PB, Friml J (2007) Molecular and cellular aspects of auxin-transport-mediated development. Trends Plant Sci 12:160–168
Vlachonasios KE, Thomashow MF, Triezenberg SJ (2003) Disruption mutations of ADA2b and GCN5 transcriptional adaptor genes dramatically affect Arabidopsis growth, development, and gene expression. Plant Cell 15:626–638

Vogel JP, Schuerman P, Woeste K, Brandstatter I, Kieber JJ (1998a) Isolation and characterization of Arabidopsis mutants defective in the induction of ethylene biosynthesis by cytokinin. Genetics 149:417–427
Vogel JP, Woeste KE, Theologis A, Kieber JJ (1998b) Recessive and dominant mutations in the ethylene biosynthetic gene ACS5 of Arabidopsis confer cytokinin insensitivity and ethylene overproduction, respectively. Proc Natl Acad Sci 95:4766–4771
Wang L et al (2017) Ethylene induces combinatorial effects of histone H3 acetylation in gene expression in Arabidopsis. BMC Genomics 18:538
Wang L, Wu J (2013) The essential role of jasmonic acid in plant–herbivore interactions–using the wild tobacco Nicotiana attenuata as a model. J Genet Genomics 40:597–606
Wang X, Chory J (2006) Brassinosteroids regulate dissociation of BKI1, a negative regulator of BRI1 signaling, from the plasma membrane. Science 313:1118–1122
Wani SH, Kumar V, Shriram V, Sah SK (2016) Phytohormones and their metabolic engineering for abiotic stress tolerance in crop plants. The Crop Journal 4:162–176
Wasternack C, Hause B (2013) Jasmonates: biosynthesis, perception, signal transduction and action in plant stress response, growth and development. Ann Bot 111:1021–1058
Xia J, Zhao H, Liu W, Li L, He Y (2009a) Role of cytokinin and salicylic acid in plant growth at low temperatures. Plant Growth Regul 57:211
Xia X-J et al (2009b) Reactive oxygen species are involved in brassinosteroid-induced stress tolerance in cucumber. Plant Physiol 150:801–814
Xie X, Yoneyama K, Yoneyama K (2010) The strigolactone story. Annu Rev Phytopathol 48:93–117
Xie Y et al (2018) An atypical R2R3 MYB transcription factor increases cold hardiness by CBF-dependent and CBF-independent pathways in apple. New Phytol 218:201–218
Yamada M, Greenham K, Prigge MJ, Jensen PJ, Estelle M (2009) The TRANSPORT INHIBITOR RESPONSE2 gene is required for auxin synthesis and diverse aspects of plant development. Plant Physiol 151:168–179
Yamada T, Kuroda K, Jitsuyama Y, Takezawa D, Arakawa K, Fujikawa S (2002) Roles of the plasma membrane and the cell wall in the responses of plant cells to freezing. Planta 215:770–778
Yamaguchi S (2008) Gibberellin metabolism and its regulation. Annu Rev Plant Biol 59:225–251
Yamamoto Y, Kamiya N, Morinaka Y, Matsuoka M, Sazuka T (2007) Auxin biosynthesis by the YUCCA genes in rice. Plant Physiol 143:1362–1371
Yamamuro C, Zhu J-K, Yang Z (2016) Epigenetic modifications and plant hormone action. Mol Plant 9:57–70
Yang C, Lu X, Ma B, Chen S-Y, Zhang J-S (2015) Ethylene signaling in rice and Arabidopsis: conserved and diverged aspects. Mol Plant 8:495–505
Yang Z, Cao S, Zheng Y, Jiang Y (2012) Combined salicylic acid and ultrasound treatments for reducing the chilling injury on peach fruit. J Agric Food Chem 60:1209–1212
Zhang F, Wang L, Ko EE, Shao K, Qiao H (2018) Histone deacetylases SRT1 and SRT2 interact with ENAP1 to mediate ethylene-induced transcriptional repression. Plant Cell. https://doi.org/10.1105/tpc.17.00671
Zhang F et al (2017) EIN2 mediates direct regulation of histone acetylation in the ethylene response. Proc Natl Acad Sci 114:10274–10279
Zhao Y (2010) Auxin biosynthesis and its role in plant development. Annu Rev Plant Biol 61:49–64
Zhao Y, Christensen SK, Fankhauser C, Cashman JR, Cohen JD, Weigel D, Chory J (2001) A role for flavin monooxygenase-like enzymes in auxin biosynthesis. Science 291:306–309
Zhou M, Xu M, Wu L, Shen C, Ma H, Lin J (2014) CbCBF from Capsella bursa-pastoris enhances cold tolerance and restrains growth in Nicotiana tabacum by antagonizing with gibberellin and affecting cell cycle signaling. Plant Mol Biol 85:259–275
Zhu J-Y, Sae-Seaw J, Wang Z-Y (2013) Brassinosteroid signalling. Development 140:1615–1620
Zhu J et al (2008) Involvement of Arabidopsis HOS15 in histone deacetylation and cold tolerance. Proc Natl Acad Sci 105:4945–4950
Zhu J et al (2015) Low temperature inhibits root growth by reducing auxin accumulation via ARR1/12. Plant Cell Physiol 56:727–736

Chapter 5
CBF-Dependent and CBF-Independent Transcriptional Regulation of Cold Stress Responses in Plants

N. Yahia, Shabir Hussain Wani, and Vinay Kumar

5.1 Introduction

The world's population will reach 9.1 billion by 2050, which will have to be feed. Since the 1960s, crop production has seen a very important increase through plant improvement. However, due to climate change, urbanization, as well as pollution, the total area of arable land is close to maximum utilization, with a direct effect on the increase of human undernourishment. It is imperative that other plant breeding tools in which biotechnological engineering of economically and nutritionally important traits should be critically and thoroughly examined. The major abiotic stresses worldwide causing risks to food security are high salinity, drought, submergence and extreme temperature including heat and cold (Wani and Sah 2014).

Plants are often confronted with unfavourable environmental conditions that can be called 'stress' and which results in decreased growth and yield of crops. Invariably subjected to all kinds of environmental stresses of biotic origin (aggressions by pathogens) or abiotic (excess or lack of light, abrupt fluctuations in temperature, hypoxia, water stress, salinity, etc.), plant developed varied and ingenious strategies to defend themselves and adapt to their changing environment, triggering sometimes very complex spectra of molecular, metabolic and morphogenetic responses (Fig. 5.1).

N. Yahia
Genetics and Plant Breeding Laboratory, Department of Biology,
Faculty of Sciences of Nature and Life, University of Oran, Oran, Algeria

S. H. Wani
Mountain Research Centre for Field Crops, Khudwani, Sher-e-Kashmir University of Agricultural Sciences and Technology of Kashmir, Srinagar, Jammu and Kashmir, India

V. Kumar (✉)
Department of Biotechnology, Modern College of Arts, Science and Commerce,
Savitribai Phule Pune University, Pune, India
e-mail: vinay.kumar@moderncollegegk.org

S. H. Wani, V. Herath (eds.), *Cold Tolerance in Plants*,
https://doi.org/10.1007/978-3-030-01415-5_5

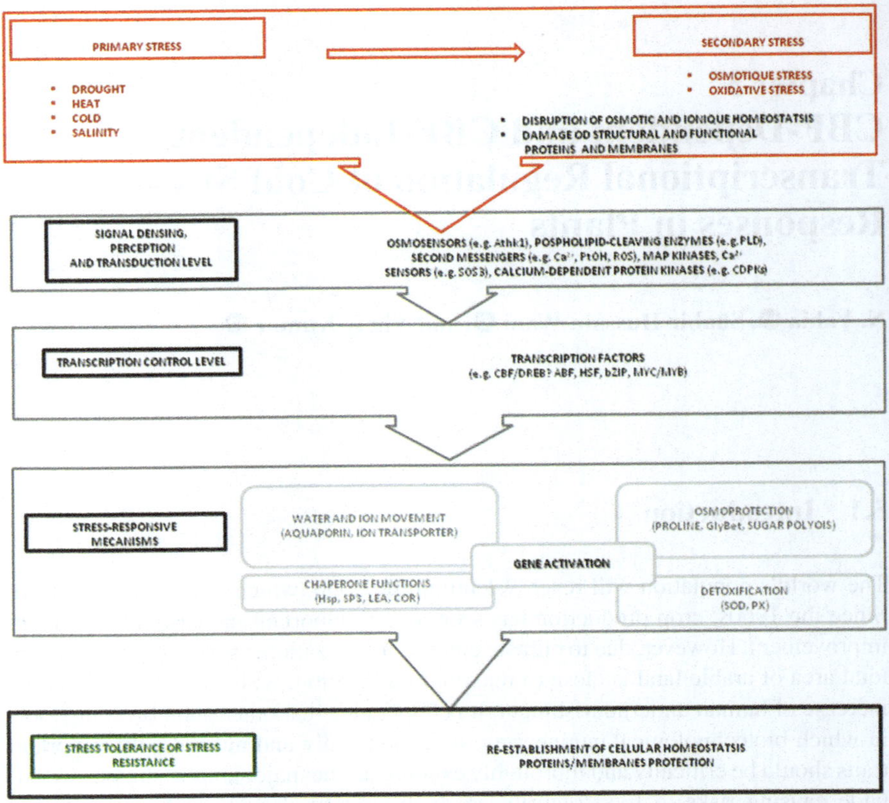

Fig. 5.1 Plant responses to abiotic stresses. Primary stresses are interrelated and provoke cellular damage as well as secondary stresses. The initial stress signal causes activation of signalling process as well as transcription control. Consequence of this is initiation of stress-responsive mechanism to restoration of cellular homeostasis, accompanied by the protection and repair of damaged proteins and membranes. Finally, plant gained tolerance or resistance to stress. ABF ABRE binding factor, Athk1 *Arabidopsis thaliana* histidine kinase-1, bZIP basic leucine zipper transcription factor, CBF/DREB C-repeat binding factor/dehydration-responsive binding protein, CDPK calcium-dependent protein kinase, COR cold-responsive protein, Hsp heat-shock protein, LEA late embryogenesis abundant, MAP mitogen-activated protein, PLD phospholipase D–PtdOH, phosphatidic acid, PX peroxidase, ROS reactive oxygen species, SOD super dismutase, SP1 stable protein 1. (Source: Gerszberg and Hnatuszko-Konka 2017)

Abiotic stresses induce, separately or in combination, general and specific adverse impacts on plant growth and development, ultimately leading to significant crop yield losses. The main abiotic stresses (drought, salinity, frost) lead to reduced water availability for essential cellular functions and maintenance of turgor pressure. However, beyond these common and well- characterized mechanisms, the response to these stresses has specificities. Most plants have developed various coping mechanisms and intricate detoxification strategies to overcome these stress conditions (Guy et al. 1985; Yahia and Fyad-Lameche 2003; Gosal et al. 2009; He et al. 2015; Yahia et al. 2015; Wani and Kumar 2015; Touati et al. 2016; Wani et al. 2016; Joshi et al. 2016).

The increased knowledge of all the metabolites formed (or metabolomics), the corresponding genes and the transcription factors (TFs) controlling their expression in response to stress conditions has bettered chances for deciphering these underlying responsive mechanisms and to target them for producing stress-tolerant plants. In the initial attempts, Guy et al. (1985) observed changes in gene expression during cold acclimatization and concluded that cold-responsive genes can achieve the biochemical and physiological changes necessary for growth and development. A large number of cold-induced genes have been isolated since then and characterized corresponding to fatty acids, chaperone proteins, proteins involved in the biosynthesis of osmoprotectants, antifreeze proteins and gene regulation components such as TFs, kinases and phosphatases, but many remain unknown. Chinnusamy et al. (2010) argued that the degree of chilling or frost tolerance of a plant therefore depends on its ability to maintain the expression of cold-inducible genes at a high level during the acclimation period.

In the initial stage of cold stress, the plants highlight changes in the expression of genes leading to synthesis of common or specific proteins. In addition, these changes in the gene expression were often accompanied by elevated levels of metabolites such as osmoprotective agents against the adverse effects of low-temperature stress (Sanghera et al. 2011). Indeed, under low temperature, multitude inducible genes have been isolated from several plants, and the expression of some of them is regulated by C-repeat binding factor/dehydration-responsive element binding (CBF/DREB1) TFs. These TFs (*bZip*, *MYC*, *MYB*, *DREB*, *NACs NAM*, *ATAF*, *CUC*, *MAP* kinase, CDP kinase, etc.) regulate gene expression, and signal transduction and function under stress responses may be useful for improving the abiotic stress tolerance in plants (Gosal et al. 2009).

Plant species react in various ways to cold stress. They modify their physiology, metabolism and growth by reprogramming gene expression. Cold stress signals in plants are transmitted to activate CBF-dependent and CBF-independent transcriptional pathways, of which the earlier one activates CBF regulon. Transcriptional cascades are next players which operate through ABA-dependent and ABA-independent pathways to induce cold-regulated (COR) gene expression, resulting in increasing levels of hundreds of metabolites, some of them being known to have protective effects against the damaging impacts of cold stress (Heidarvand and Amiri 2010). CBF TF genes are induced by the constitutively expressed ICE1 (inducer of CBF expression 1) by binding to the *CBF* promoter. ICE1–CBF cold response pathway is conserved in diverse plant species (Fig. 5.1). The cold-inducible genes are regulated mainly via transcriptional regulation. This regulation is largely modulated by proteins that bind to specific sites in the promoter regions of the genes. The CCGAC (or CRT for C-repeat) pattern forms the pattern of a low-temperature-responsive element or dehydration-responsive element (DRE). CBFs are TFs that regulate the expression of several genes in relation to abiotic stress (Chinnusamy et al. 2006; Zhao et al. 2011; Guo et al. 2011). CBF genes were also referred to as DRE-binding factor genes that code for the transcription of conserved sequences of COR (CCGAC [C-repeat (CRT)/dehydration element (DRE)]) genes at the promoter level of several genes. Responses to stress include genes for early dehydration stress responses and cold stress (Stockinger et al. 1997; Liu et al. 1998; Thomashow 2001; Shinozaki and Yamaguchi-Shinozaki 2007).

5.2 Perception and Transduction of the Cold Stress Signals

For a plant to implement effective cold tolerance mechanisms, it is necessary first to perceive the low temperatures followed by signal transmission in order to regulate the appropriate genes, and finally the proteins are synthesized to limit the damage caused by the cold (Navarro 2009). Cold stress induces immediate changes (known as primary events) in the cell balance. These primary events are the cause of the damage of the cell but also inform the cell of the existence of a stress. It is well established that the cell membrane systems are the main sites of gel injury in plants. Cell membranes are fluid structures; membrane fluidity depends on temperature but also on the lipid composition and degree of saturation of the fatty acids of the membranes (Steponkus 1984; Murata and Los 1997). The signal induced by low temperature is initially perceived by the plasma membrane either by the membrane fluidity or through membrane sensors such as Ca^{2+}, kinases, histidine kinase receptors and phospholipases. Subsequently, cytoskeletal reorganization and cytosolic Ca^{2+} influx take place. The increase of cytosolic Ca^{2+} is captured by CDPKs, phosphatases and MAPKs, whose role is signal transduction to activate the transcriptional cascade (Knight and Knight 2001; Vergnolle et al. 2005). Cold temperatures can reduce their fluidity, causing increased stiffness, and negative temperatures cause damage due to the formation of ice crystals. The presence of ice in the extracellular space reduces its water potential. This results in dehydration and contraction of the cell volume, which will destabilize the plasma membrane. When crystallization reaches the intracellular medium, it results in cell destruction and therefore tissue death (Guy 1990).

The major difference between acclimated and non-acclimated plant membranes is that the membrane material remains intact during the freeze-thaw cycle in the tolerant cells. The cells of the tolerant plants are able to modify their wall and the plasma membrane to protect the plant from injury caused by freezing. Indeed, the protoplasts of acclimated plants do not form an endocytotic vesicle that would inevitably lead to a loss of surface. On the other hand, exocytotic extrusions are formed that allow the membrane to recover its original surface without tearing during the thaw period. Freezing or frost tolerance, therefore, was closely related to the mechanisms by which plant cells avoid injury to their cellular membranes. Several studies have shown that, during cold stress, the levels of sterols and phospholipids in the plasma membrane increase. On the other hand, the levels of acetylated glucosides and cerebrosides (CER) decrease. These changes were observed in *Arabidopsis*, rye and oats during cold acclimatization. These changes would have a role in the cryostability of the plasma membrane during freezing (Uemura et al. 1995, 2006; Yamazaki et al. 2009). In the similar vein, Zhu (2016) reported that the mechanisms of cold-sensing in plants are that cold shock leads to changes in membrane fluidity and rearrangement of the cytoskeleton that trigger Ca^{2+} influx; consequently it activates the induction of COR genes.

Ma et al. (2015) showed that COLD1 gene plays a key role in enhancing chilling tolerance in rice line. It stimulates intracellular Ca^{2+} influx by associating with the α-subunit of the G protein and accelerates the activity of the G-GTPase protein, to ultimately up-regulate the expression of the COR genes, under cold stress. Furthermore, abscisic acid (ABA) and reactive oxygen species (ROS) can also induce Ca^{2+} signatures that influence cold signalling (Chinnusamy et al. 2007). If the input of Ca^{2+} ions is inhibited, the low-temperature tolerance decreases (Sung et al. 2003).

Among the main molecules involved in signal transduction are ABA (Leung and Giraudat 1998) and ROS (Lamb and Dixon 1997). ABA is a stress hormone that regulates many aspects of plant development such as seed germination, dormancy and tolerance to seed desiccation, but it also plays a major role in the response to biotic and abiotic stresses (Chinnusamy et al. 2004). ABA improves antioxidant defence and slows ROS accumulation caused by low temperatures (Liu et al. 2011). Thus, it regulates the expression of many genes involved in resistance to cold, osmotic shock or dehydration (Xiong et al. 2001).

Some symptoms observed under stress of biotic or abiotic origin are the consequence of a strong accumulation of oxygen free radicals and an alteration of cellular homeostasis. Plants exposed to low temperatures produce reactive forms of oxygen that damage membrane lipids, proteins, chlorophyll and nucleic acids. Although reactive forms of oxygen are formed during the normal metabolism of the plant, increasing their intracellular concentration is often synonymous with stress. In addition to their effect on calcium signatures, they are also involved in the signalling cascades responsible for the induction and regulation of many defence genes. Thus, ROS serves as a second messenger for activation of stress response and defence mechanisms (Mittler 2002; Apel and Hirt 2004).

5.3 Regulation of Gene Expression in Response to Low Temperatures

The expression of gene cascades in plants is regulated by abiotic environmental stresses including cold (Thomashow 1999; Shinozaki et al. 2003; Shi et al. 2018a, b). Sensitive or cold tolerant, acclimated or not acclimated, all plants are generally capable of experiencing low temperatures and modifying gene expression in response to these conditions, though with a varying degree. Large modifications at the transcriptome scale are noted each time (Carvallo et al. 2011; Maruyama et al. 2012), leading to modifications of the proteome and the metabolome. Recently, Kumar and Wigge (2010) showed that temperature can have a direct effect on chromatin remodelling and thus on the activation of transcription of many genes.

Many abiotic inducible genes are controlled by ABA, but not all, indicating that both ABA-dependent and independent ABA regulation systems are involved in the expression of stress-sensitive genes (Thomashow 1999; Shinozaki et al. 2003). In response to cold stress, the independent ABA pathway leads to the activation of a family of TF genes called DREBs or CBFs, in turn inducing effector genes by binding to the cis CRT/DRE element (C-repeat/dehydration element) located in their promoter region (Stockinger et al. 1997). At the same time, the cold induces an accumulation of the ABA hormone, which leads to the expression of the ABA-responsive element binding factor which can also induce the effector genes but via another cis sequence: the ABA-responsive element (ABRE) (Liu et al. 1998).

5.3.1 CBF Pathway

CBF/DREB1 genes are key players in the control of the cold acclimatization process in plants (Thomashow 2010; Shi et al. 2018a, b). They regulate the expression of a set of target genes that are involved in transcription by binding to the cis element LTRE/CRT/DRE (low-temperature-responsive element/C-repeat/dehydration element) located on their promoter sequence. This set of genes was generally referred to as the CBF regulon. Thus, CBFs regulate the expression of genes that are involved in phosphoinositide metabolism, osmolyte biosynthesis, ROS detoxification, membrane transport, hormone metabolism, signalling and many others with known or suspected cellular protection (Lee et al. 2005; Fowler and Thomashow 2002; Maruyama et al. 2004). CBFs appear to be involved in growth control, leaf surface area, cell volume, leaf thickness, stomatal density and anthocyanin synthesis (Gilmour et al. 2004; Savitch et al. 2005; Pino et al. 2008). Overexpression of AtCBF1 and AtCBF3 enhanced frost tolerance in non-acclimated plants and induced soluble sugars and proline accumulated in plants under warm conditions (Jaglo-Ottosen et al. 1998).

The number of CBF/DREB1 genes varies from one species to another but also at the intraspecific scale. Generally, it was considered that the genomes of monocotyledons (especially Triticeae) have more CBF/DREB1 genes than those of dicotyledons. In monocotyledonous plants such as barley and wheat, the family has up to 25 members (Badawi et al. 2007), whereas in dicotyledonous plants, the literature reports a maximum of six sequences as in *Arabidopsis* and poplar (Haake et al. 2002; Benedict et al. 2006). CBF genes are highly conserved between species and even more so between members of the same family as is the case in the vine or the *V. vinifera*. CBF4 sequence is 99% identical to that of *V. riparia* (Xiao et al. 2008).

CBF/DREB1 genes generally respond to cold stress in cold-acclimatizing monocot plants as demonstrated in wheat, barley (Choi et al. 2002; Marozsán-Tóth et al. 2015), ryegrass (Xiong and Fei 2006), rice (Dubouzet et al. 2003), maize (Qin et al. 2004) or dicotyledons such as rapeseed (Gao et al. 2002) or tomato (Jaglo et al. 2001). Similarly, the results showed that the EguCBF1a, EguCBF1b, EguCBF1c and EguCBF1d genes isolated from *E. gunnii* respond strongly to cold shocks in

E. gunnii and *E. gunnii* × *E. dalrympleana* but also for species more sensitive to cold as *E. urophylla* × *E. grandis*.

Since the *Arabidopsis* genome sequenced, six CBF/DREB1 genes have been isolated. The initial investigations on *Arabidopsis* were focused on cold-induced CBF1, 2 and 3, arranged in tandem with chromosome 4 (Gilmour et al. 1998; Medina et al. 1999). Isolated AtCBF4 gene was induced by drought and is located on chromosome 5 (Haake et al. 2002). On the other hand, the CBF5 and 6 genes, named AtDDF1 and AtDDF2 induced by salt stress, were mainly located on chromosome 1 (Magome et al. 2004). The CBF pathway accounts for 5–15% of the transcript changes in response to cold in *Arabidopsis* (Hannah et al. 2006). DNA sequencing and result mapping indicated that the CBF1, CBF2 and CBF3 genes are present in the direct repeat genome, CBF1-CBF3-CBF2, on chromosome 4 at 72.8 cM, closely related to molecular markers PG11 and m600 (Gilmour et al. 1998).

M. truncatula has at least 17 CBF/DREB1 genes (considering MtCBF2a and MtCBF2b) all located in homologous regions (Azar et al. 2011). In cereals, recent studies show that CBF2A in barley and CBF14 and CBF15 in wheat have been directly proven involved in cold acclimatization causing a significant increase in frost tolerance (Soltész et al. 2013; Jeknic et al. 2014).

In woody plants also, CBF genes are identified; the first CBF of isolated woody cherry (*Prunus avium*) is designated PaDREB1 or PaD2B (Kitashiba et al. 2002). EguCBF1a and EguCBF1b are the first isolated and characterized CBF in *Eucalyptus* (*E. gunnii*). They are strongly regulated by the cold, but not so by other abiotic stresses (El Kayal et al. 2006). CBFs were then isolated from other trees such as poplar, apple, birch and grapevine. Among the six CBF genes isolated from *Populus trichocarpa*, only PtCBF1 and PtCBF2 were phylogenetically similar to *Arabidopsis* AtCBF1-4. These genes respond to cold stress and were differently expressed in different plant parts (stems and leaves, Benedict et al. 2006). In addition, four CBFs isolated from birch (*Betula pendula*) respond to cold and exhibit rapid and transient expression in active tissues in long daylight and late, more durable expression in short days in dormant tissues. This suggests the involvement of CBFs in the winter acclimatization of this woody species (Welling and Palva 2008).

Xiao et al. (2006) showed that four CBF are common in *Vitis vinifera* (cold sensitive) and *V. riparia* (cold tolerant). VvCBF1-3 and VrCBF1-3 are induced by cold, dryness and ABA, while the more distinct VvCBF4 and VrCBF4 sequence respond for at least 1 day and only identical way in both species (Xiao et al. 2008). CBF have also been identified in tropical deciduous plants such as *Hevea* (*Hevea brasiliensis*). In tropical evergreen plants closest to *E. gunnii*, a cold-responsive EgCBF1 was isolated from *E. globulus* (Gamboa et al. 2007). Houde et al. (2004), based on the ion leakage experiments, concluded that Wcor410a acidic dehydrin gene from wheat transferred in transgenic strawberry, conferred frost tolerance, and wcor410 protein was expressed in transgenic strawberry at the level comparable with that in cold-acclimated wheat. Further, two cold-induced CBFs were isolated from the tolerant species *Poncirus trifoliata* and the susceptible species *Citrus paradisi* (Champ et al. 2007).

5.3.2 *CBF: The Independent Pathway*

It is conventionally established that phytohormones such as auxins, cytokinins, ethylene, gibberellins and abscisic acid ABA, in addition to new members, brassinosteroids, jasmonates and strigolactones as it happens, could prove to be potential targets for their engineering for producing abiotic stress tolerance crop plants (Wani et al. 2016). Among these phytohormones, ABA is the most studied in the engineering of abiotic stress tolerance in plants of agronomic interest because of its importance as a stress hormone and for its preponderant role and extensive functions related to environmental stresses (Wani and Kumar 2015). Phytohormones including auxin, cytokinins, ABA, gibberellins, jasmonic acid, ethylene and brassinosteroids are intimately linked in the CBF-independent transcriptional pathway components under cold acclimation (Zhao et al. 2014; Wani et al. 2016; Joshi et al. 2016).

The CRT/DRE motif is a 5 bp CBF/DREB1 binding motif, which forms the conserved central element of a cis-acting regulatory element, in which the CBF TFs bind (Yamaguchi-Shinozaki and Shinozaki, 1994; Stockinger et al. 1997; Liu et al. 1998). Agarwal et al. (2006) showed that DREB proteins are important TFs that induce a set of abiotic stress-related genes to confer stress resistance in the plants. The DREB TFs can be divided in two types, DREB1 and DREB2, which control two different signal transduction pathways, the first for low- temperature stress and the second for dehydration stress. Clusters of CBF/DREB1 genes repeated in tandem have been demonstrated in several species, other than *Arabidopsis* (6 genes) (Gilmour et al. 1998), such as *Triticum monococcum* (14 genes) (Miller et al. 2006) and *Hordeum vulgare* (12 genes) (Skinner et al. 2006). Interestingly, even the ectopic expression of CBFs is reported sufficient to activate the expression of COR genes and induce cold acclimation, even at warm temperatures (Stockinger et al. 1997; Liu et al. 1998; Patel and Franklin 2009; Zhao et al. 2011; Shi et al. 2015; Erimina et al. 2016; Gerszberg and Hnatuszko-Konka 2017).

The gene products of the AP2/EREBP family are characterized by a conserved domain of DNA binding (Okamuro et al. 1997; Zhao et al. 2014) which is of the order of 60 amino acids: the AP2/ERF domain which is the domain of DNA binding (Jofuku et al. 1994; Ohme-Takagi and Shinshi 1995). The AP2/ERF domain also presumably contains the address signal to the kernel (Canella et al. 2010). The CBF/DREB1 genes have a simple structure with an average 700 bp coding region not including introns. Recent report by Erimina et al. (2016) showed that it is clear that hormones act as central regulators of cold stress responses in plant. Shi et al. (2015) reported that hormonal components play important roles in regulating plant freezing tolerance by either CBF-dependent or CBF-independent pathways.

5.4 Conclusion

Stress conditions induce signalling reactions that can lead to the establishment of defences or trigger programmed cell death. Cold induce osmotic stress and oxidative stress, which ultimately leads to an imbalance of homeostasis and weakening of

Table 5.1 Examples of genes exploited for enhanced tolerance to cold stress

Plant	Gene	Effect of gene engineering	References
Pyrus betulaefolia	*PbrMYB5*	PbrMYB5 overexpression contributes to the improvement of cold tolerance in transgenic plant	Xing et al. (2018)
Camellia sinensis	*KCSs, NAC080, SWEETs, and ENOs*	Regulates cold tolerance in tea plant leaves	Hao et al. (2018)
Arabidopsis thaliana	*AtMYB14*	Regulates cold tolerance in *Arabidopsis*	Chen et al. (2013)
A. thaliana	*AtMYB14*	AtMYB14 participates in freezing tolerance in *Arabidopsis* by affecting expression of CBF genes. Encodes a nuclear protein that functions as an R2R3-MYB transcription activator	Chen et al. (2013)
Dendranthema grandiflorum	*DEG genes*	Cold-responsive genes related to low-temperature sensing and signal transduction, membrane lipid stability, ROS scavenging and osmoregulation	Wang et al. (2018)
Lycopersicon esculentum	*Osmotin*	Enhanced tolerance to cold	Patade et al. (2013)
Oryza sativa	*OsNAC*	OsNAC regulates the expression NACRS target gene to increase cold tolerance	Zang et al. (2017)
Medicago falcata	*MfTIL1*	MfTIL1 confer elevated survival rate in response to freezing in transgenic tobacco plants	He et al. (2015)
Oryza sativa	OsmiR156k	OsmiR156k is suggested a negative regulator of plant tolerance to cold stress	Cui et al. (2015)
Solanum melongena L.	*POD and CAT genes*	POD and CAT relative gene expression enhances chilling tolerance and reduces damage in cold-stored eggplant	Shi et al. (2018a, b)
Artemisia annua	*Aa547*	Aa 547 peroxidase gene overexpression under cold stress	Nair et al. (2018)
Medicago truncatula	*CBF/DREB1*	CBF/DREB1 genes located in a major freezing tolerance QTL region on *Medicago truncatula* chromosome 6	Tayeh et al. (2013)
Oryza sativa	*COLD1*	Transmembrane protein regulates GTPase activity under chilling stress	Ma et al. (2015)

membranes and proteins. The perception and the transduction of the signal leads to the altered expression of regulatory genes which in turn controls the expression patterns of effector genes allowing the establishment of stress responses and tolerance mechanisms. During cold acclimation, several molecular and physiological factors are involved and are crucial. The TF CBFs/DREBs play crucial roles in the regulation of genes involved in tolerance to low-temperature stress; once the plants sense low temperature, they activate a complex cold acclimation process and activation of CBF-dependent and CBF-independent pathways. With regard to CBF-independent, more than 80% components of cold-responsive genes are not controlled directly by

CBFs. Transcriptome analyses have indicated that only ~12% of the cold-responsive genes are controlled by CBFs. During cold stress, a cascade of phosphorylation triggered by the influx of cytosolic Ca^{2+} and the induction of CBF genes gets underway. In the future, it is interesting to focus on studies targeting the role of phytohormones, small RNAs, as well as the effects of overexpression of CBF genes for the improvement of agronomic species and in order to better understand the cold stress response mechanisms (Table 5.1).

References

Agarwal M, Hao Y, Kapoor A et al (2006) A R2R3 type MYB transcription factor is involved in the cold regulation of CBF genes and in acquired freezing tolerance. J Biol Chem 281:37636–37645

Apel K, Hirt H (2004) Reactive oxygen species: metabolism, oxidative stress and signal transduction. Annu Rev Plant Biol 55:373–399

Azar S, San-Clemente H, Marque G et al (2011) Bioinformatic prediction of the AP2/ERF family genes in *Eucalyptus grandis*: focus on the CBF family. BMC Proc 5(Suppl7):165

Badawi M, Danyluk J, Boucho B et al (2007) The CBF gene family in hexaploid wheat and its relationship to the phylogenetic complexity of cereal CBFs. Mol Gen Genomics 277:533–554

Benedict C, Skinner JS, Meng R (2006) The CBF1-dependent low temperature-signalling pathway, regulon and increase in freeze tolerance are conserved in *Populus spp*. Plant Cell Environ 29:1259–1272

Canella D, Gilmour SJ, Kuhn LA et al (2010) DNA binding by the *Arabidopsis* CBF1 transcription factor requires the PKKP/RAGRxKFxETRHP signature sequence. Biochem Biophys Acta 1799:454–462

Carvallo MA, Pino MT, Jeknić Z et al (2011) A comparison of the low temperature transcriptome and CBF regulon of three plant species that differ in freezing tolerance: *Solanum commersonii*, *Solanum tuberosum*, and *Arabidopsis thaliana*. J Exp Bot 62:3807–3819

Champ KI, Febres VJ, Moore GA (2007) The role of CBF transcriptional activators in two *Citrus* species (*Poncirus* and *Citrus*) with contrasting levels of freezing tolerance. Physiol Plant 129:529–541

Chen Y, Chen Z, Kang J et al (2013) AtMYB14 regulates cold tolerance in *Arabidopsis*. Plant Mol Biol Report 31:87–97

Chinnusamy V, Schumaker K, Zhu JK (2004) Molecular genetics perspectives on cross-talk and specificity in abiotic stress signalling in plants. J Exp Bot 55:225–236

Chinnusamy V, Zhu J, Zhu JK (2006) Gene regulation during cold acclimation in plants. Physiol Plant 126:52–61

Chinnusamy V, Zhu J, Zhu JK (2007) Cold stress regulation of gene expression in plants. Trends Plant Sci 12:444–451

Chinnusamy V, Zhu JK, Sunkar K (2010) Gene regulation during cold stress acclimation in plants. Methods Mol Biol 639:39–55

Choi DW, Rodriguez EM, Close TJ (2002) Barley Cbf3 gene identification, expression pattern, and map location. Plant Physiol 129:1781–1787

Cui N, Sun X, Sun M et al (2015) Overexpression of OsmiR156k leads to reduced tolerance to cold stress in rice (Oryza Sativa). Mol Breed 35:214. https://doi.org/10.1007/s11032-015-0402-6

Dubouzet JG, Sakuma Y, Ito Y et al (2003) OsDREB genes in rice, Oryza sativa L., encode transcription activators that function in drought-, high-salt- and cold-responsive gene expression. Plant J 33:751–763

Erimina M, Rozhon W, Poppenberger B (2016) Hormonal control of cold stress responses in plants. Cell Mol Life Sci 73:797–810

Fowler S, Thomashow MF (2002) Arabidopsis transcriptome profiling indicates that multiple regulatory pathways are activated during cold acclimation in addition to the CBF cold response pathway. Plant Cell 14:1675–1690

Gamboa MC, Rasmussen-Poblete S, Valenzuela PD et al (2007) Isolation and characterization of a cDNA encoding a CBF transcription factor from *E. globulus*. Plant Physiol Biochem 45:1–5

Gao MJ, Allard G, Byass L et al (2002) Regulation and characterization of four CBF transcription factors from *Brassica napus*. Plant Mol Biol 49:459–471

Gerszberg A, Hnatuszko-Konka K (2017) Tomato tolerance to abiotic stress: a review of most often engineered target sequences. Plant Growth Regul 83:175–198

Gilmour SJ, Zarka DG, Stockinger EJ et al (1998) Low temperature regulation of the *Arabidopsis* CBF family of AP2 transcriptional activators as an early step in cold-induced COR gene expression. Plant J 16:433–442

Gilmour SJ, Fowler SG, Thomashow MF (2004) Arabidopsis transcriptional activators CBF1, CBF2, and CBF3 have matching functional activities. Plant Mol Biol 54:767–781

Gosal SS, Wani SH, Manjit SK (2009) Biotechnology and drought tolerance. J Crop Improv 23:19–54

Guo HM, Li ZC, Zhang H (2011) Cloning of cotton CBF gene for cold tolerance and its expression in transgenic tobacco. Acta Agron Sin 37:286–293

Guy C (1990) Cold acclimation and freezing stress tolerance: role of protein metabolism. Annu Rev Plant Physiol Plant Mol Biol 41:187–223

Guy CL, Niemi KJ, Bramble R (1985) Altered gene expression during cold acclimation of spinach. Proc Natl Acad Sci U S A 82:3673–3677

Haake V, Cook D, Riechmann JL et al (2002) Transcription factor CBF4 is a regulator of drought adaptation in *Arabidopsis*. Plant Physiol 130:639–648

Hannah MA, Wiese D, Freund S et al (2006) Natural genetic variation of freezing tolerance in *Arabidopsis*. Plant Physiol 142:98–112

Hao X, Wang B, Wang L et al (2018) Comprehensive transcriptome analysis reveals common and specific genes and pathways involved in cold acclimation and cold stress in tea plant leaves. Sci Hortic 240:354–368

He X, Sambe MAN, Zhuo C et al (2015) A temperature induced lipocalin gene from *Medicago falcata* (*MfTIL1*) confers tolerance to cold and oxidative stress. Plant Mol Biol 87:645–654

Heidarvand L, Amiri LM (2010) What happens in plant molecular responses to cold stress? Acta Physiol Plant 32:419–431

Houde M, Dallaire S, N'Dong D et al (2004) Overexpression of the acidic dehydrin WCOR410 improves freezing tolerance in transgenic strawberry leaves. Plant Biotechnol J 2:381–387

Jaglo KR, Kleff S, Amundsen KL et al (2001) Components of the *Arabidopsis* C-repeat/dehydration-responsive element binding factor cold-response pathway are conserved in *Brassica napus* and other plant species. Plant Physiol 127:910–917

Jaglo-Ottosen KR, Glimour SJ, Zarka DG et al (1998) Arabidopsis CBF 1 overexpression induced COR genes and enhances freezing tolerance. Science 280:104–106

Jeknic Z, Pillman KA, Dhillon T et al (2014) Hv-CBF2A overexpression in barley accelerates COR gene transcript accumulation and acquisition of freezing tolerance during cold acclimation. Plant Mol Biol 84:67–82

Jofuku KD, den-Boer BG, Van Montagu M et al (1994) Control of *Arabidopsis* flower and seed development by the homeotic gene APETALA2. Plant Cell 6:1211–1225

Joshi R, Wani SH, Singh B et al (2016) Transcription factors and plants response to drought stress: current understanding and future directions. Front Plant Sci 7:1029. https://doi.org/10.3389/fpls.2016.01029

Kayal EW, Navarro M, Marque G et al (2006) Expression profile of CBF-like transcriptional factor genes from *Eucalyptus* in response to cold. J Exp Bot 57(10):2455–2469

Kitashiba H, Matsuda N, Ishizaka T et al (2002) Isolation of genes similar to DREB1/CBF from sweet cherry (*Prunus avium* L.). J Jap Soc Hort Sci 71:651–657

Knight H, Knight MR (2001) Abiotic stress signaling pathways: specificity and crosstalk. Trends Plant Sci 6:262–267

Kumar SV, Wigge PA (2010) H2A.Z-containing nucleosomes mediate the thermosensory response in *Arabidopsis*. Cell 140:136–147

Lamb CJ, Dixon RA (1997) The oxidative burst in plant disease resistance. Annu Rev Plant Physiol Plant Mol Biol 48:251–275

Lee BH, Henderson DA, Zhu JK (2005) The *Arabidopsis* cold-responsive transcriptome and its regulation by ICE1. Plant Cell 17:3155–3175

Leung J, Giraudat J (1998) Abscisic acid signal transduction. Annu Rev Plant Biol 49:199–222

Liu Q, Kasuga M, Sakuma Y et al (1998) Two transcription factors, DREB1 and DREB2, with an EREBP/AP2 DNA binding domain separate two cellular signal transduction pathways in drought- and low-temperature-responsive gene expression, respectively, in *Arabidopsis*. Plant Cell 10:1391–1406

Liu Y, Jiang H, Zhao Z et al (2011) Abscisic acid is involved in brassinosteroids-induced chilling tolerance in the suspension cultured cells from *Chorispora bungeana*. J Plant Physiol 168:853–862

Ma Y, Dai X, Xu Y et al (2015) COLD1 confers chilling tolerance in rice. Cell 160:1209–1221

Magome H, Yamaguchi S, Hanada A et al (2004) *dwarf* and *delayed flowering 1*, a novel *Arabidopsis* mutant deficient in gibberellin biosynthesis because of overexpression of a putative AP2 transcription factor. Plant J 37:720–729

Marozsan-Toth Z, Vashegyi I, Galiba G et al (2015) The cold response of CBF genes in barley is regulated by distinct signaling mechanisms. J Plant Physiol 181:42–49

Maruyama K, Sakuma Y, Kasuga M et al (2004) Identification of cold-inducible downstream genes of the *Arabidopsis* DREB1A/CBF3 transcriptional factor using two microarray systems. Plant J 38:982–993

Maruyama K, Todaka D, Mizoi J et al (2012) Identification of cis-acting promoter elements in cold- and dehydration- induced transcriptional pathways in *Arabidopsis*, rice, and soybean. DNA Res 19:37–49

Medina J, Bargues M, Terol J et al (1999) The *Arabidopsis* CBF gene family is composed of three genes encoding AP2 domain-containing proteins whose expression is regulated by low temperature but not by abscisic acid or dehydration. Plant Physiol 119:463–469

Miller AK, Galiba G, Dubcovsky J (2006) A cluster of 11 CBF transcription factors is located at the frost tolerance locus Fr-A (m) 2 in *Triticum monococcum*. Mol Gen Genomics 275:193–203

Mittler R (2002) Oxidative stress, antioxidants and stress tolerance. Trends Plant Sci 7:405–410

Murata N, Los A (1997) Membrane fluidity and temperature perception. Plant Physiol 115:875–879

Nair P, Shasany AK, Khan F et al (2018) Differentially expressed peroxidases from *Artemisia annua* and their responses to various abiotic stresses. Plant Mol Biol Report 36:295. https://doi.org/10.1007/s11105-018-1078-y

Navarro M (2009) Etude fonctionnelle de gènes de facteurs de transcription CBFs impliqués dans la tolérance au froid chez l'*Eucalyptus*. In: Thèse de doctorat. Université Toulouse III – Paul Sabatier, p 155

Ohme-Takagi M, Shinshi H (1995) Ethylene-inducible DNA binding proteins that interact with an ethylene-responsive element. Plant Cell 7:173–182

Okamuro JK, Caster B, Villarroel R et al (1997) The AP2 domain of APETALA2 defines a large new family of DNA binding proteins in Arabidopsis. Proc Natl Acad Sci U S A 94:7076–7081

Patade VY, Khatri D, Kumari M et al (2013) Cold tolerance in osmotin transgenic tomato (*Solanum lycopersicum* L.) is associated with modulation in transcript abundance of stress responsive genes. Springerplus 2:117. https://doi.org/10.1186/2193-1801-2-117

Patel D, Franklin KA (2009) Temperature-regulation of plant architecture. Plant Signal Behav 4:577–579

Pino MT, Skinner JS, Jeknic Z et al (2008) Ectopic AtCBF1over-expression enhances freezing tolerance and induces cold acclimation-associated physiological modifications in potato. Plant Cell Environ 31:393–406

Qin F, Sakuma Y, Li J et al (2004) Cloning and functional analysis of a novel DREB1/CBF transcription factor involved in cold-responsive gene expression in *Zea mays* L. Plant Cell Physiol 45:1042–1052

Sanghera GS, Wani SH, Hussain W et al (2011) Engineering cold stress tolerance in crop plants. Curr Genomics 12:30–43

Savitch LV, Allard G, Seki M et al (2005) The effect of over-expression of two *Brassica* CBF/DREB1-like transcription factors on photosynthetic capacity and freezing tolerance in *Brassica napus*. Plant Cell Physiol 46:1525–1539

Shi Y, Ding Y, Yang S (2015) Cold signal transduction and its interplay with phytohormones during cold acclimation. Plant Cell Physiol 56:7–15

Shi J, Zuoa J, Zhoua F et al (2018a) Low-temperature conditioning enhances chilling tolerance and reduces damage in cold-stored eggplant (*Solanumm elongena* L.) fruit. Postharvest Biol Technol 141:33–38

Shi Y, Ding Y, Yang S (2018b) Molecular regulation of CBF signaling in cold acclimation. Trends Plant Sci 23:623–637

Shinozaki K, Yamaguchi-Shinozaki K (2007) Gene networks involved in drought stress response and tolerance. J Exp Bot 58:221–227

Shinozaki K, Yamaguchi-Shinozaki K, Seki M (2003) Regulatory network of gene expression in the drought and cold stress responses. Curr Opin Plant Biol 6:410–417

Skinner JS, Szűcs P, Von Zitzewitz J et al (2006) Mapping of barley homologs to genes that regulate low temperature tolerance in *Arabidopsis*. Theor Appl Genet 112:832–842

Soltész A, Smedley M, Vashegyi I et al (2013) Transgenic barley lines prove the involvement of TaCBF14 and TaCBF15 in the cold acclimation process and in frost tolerance. J Exp Bot 64:1849–1862

Steponkus PL (1984) Role of the plasma membrane in freezing injury and cold acclimation. Annu Rev Plant Physiol 35:543–584

Stockinger EJ, Gilmour SJ, Thomashow MF (1997) *Arabidopsis thaliana* CBF1encodes an AP2 domain-containing transcriptional activator that binds to the C-repeat/DRE, a cis-acting DNA regulatory element that stimulates transcription in response to low temperature and water deficit. Proc Natl Acad Sci U S A 94:1035–1040

Sung DY, Kaplan F, Lee KJ et al (2003) Acquired tolerance to temperature extremes. Trends Plant Sci 8:179–187

Tayeh N, Bahman N, Sellier H et al (2013) A tandem array of CBF/DREB1 genes is located in a major freezing tolerance QTL region on Medicago truncatula chromosome 6. BMC Genomics 14:814. https://doi.org/10.1186/1471-2164-14-814

Thomashow MF (1999) Plant cold acclimation: freezing tolerance genes and regulatory mechanisms. Annu Rev Plant Physiol Plant Mol Biol 50:571–599

Thomashow MF (2001) So what's new in the field of plant cold acclimation? Lots! Plant Physiol 125:89–93

Thomashow MF (2010) Molecular basis of plant cold acclimation: insights gained from studying the CBF cold response pathway. Plant Physiol 154:571–577

Touati A, Yahia N, Fyad-Lamèche FZ (2016) Morphometric variability and biochemical analysis of growth seedlings under salt stress in tomato (*Lycopersicon esculentum* Mill.) cultivars. Mol Plant Breed 7:1–9

Uemura M, Raymond AJ, Steponkus PL (1995) Cold acclimation of *Arabidopsis thaliana*: effect on plasma membrane lipid composition and freeze induced lesions. Plant Physiol 109:15–30

Uemura M, Tominaga Y, Nakagawara C et al (2006) Responses of the plasma membrane to low temperatures. Physiol Plant 126:81–89

Vergnolle C, Vaultier MN, Taconnat L et al (2005) The cold-induced early activation of phospholipase C and D pathways determines the response of two distinct clusters of genes in *Arabidopsis* cell suspensions. Plant Physiol 139:1217–1233

Wang K, Bai ZY, Liang QY et al (2018) Transcriptome analysis of chrysanthemum (*Dendranthema grandiflorum*) in response to low temperature stress. BMC Genomics 19:319. https://doi.org/10.1186/s12864-018-4706-x

Wani SH, Kumar V (2015) Plant stress tolerance: engineering ABA: a potent Phytohormone. Transcriptomics 3:1000113. https://doi.org/10.4172/2329-8936.1000113

Wani SH, Sah SK (2014) Biotechnology and abiotic stress tolerance in rice. J Rice Res 2:e105. https://doi.org/10.4172/jrr.1000e105

Wani SH, Kumar V, Shriram V et al (2016) Phytohormones and their metabolic engineering for abiotic stress tolerance in crop plants. Crop J 4:162–176

Welling A, Palva ET (2008) Involvement of CBF transcription factors in winter hardiness in birch. Plant Physiol 147:1199–1211

Xiao H, Siddiqua M, Braybrook S et al (2006) Three grape *CBF/DREB1* genes respond to low temperature, drought and abscisic acid. Plant Cell Environ 29:1410–1421

Xiao H, Tattersall EAR, Siddiqua MK et al (2008) CBF4 is a unique member of the CBF transcription factor family of *Vitis vinifera* and *Vitis riparia*. Plant Cell Environ 31:1–10

Xing C, Liu Y, Zhao L et al (2018) A novel MYB transcription factor regulates AsA synthesis and effects cold tolerance. Plant Cell Environ. https://doi.org/10.1111/pce.13387

Xiong Y, Fei SZ (2006) Functional and phylogenetic analysis of a DREB/CBF-like gene in perennial ryegrass (*Lolium perenne* L.). Planta 224:878–888

Xiong L, Lee B, Ishitani M et al (2001) FIERY1 encoding an inositol polyphosphate 1-phosphatase is a negative regulator of abscisic acid and stress signalling in *Arabidopsis*. Genes Dev 15:1971–1984

Yahia N, Fyad-Lamèche FZ (2003) Evaluation de la variabilité de jeunes plants de *Medicago* soumis à un régime de basse température. Acta Bot Gallica 150:3–17

Yahia N, Ali N, Habouche D et al (2015) Changes of peroxidase activities under cold stress in annuals populations of *Medicago*. Mol Plant Breed 6:1–9

Yamaguchi-Shinozaki K, Shinozaki K (1994) A novel cis-acting element in an *Arabidopsis* gene is involved in responsiveness to drought, low-temperature, or high-salt stress. Plant Cell 6:251–264

Yamazaki T, Kawamura Y, Uemura M (2009) Extracellular freezing-induced mechanical stress and surface area regulation on the plasma membrane in cold-acclimated plant cells. Plant Signal Behav 4:231–233

Zang Q, Chen Q, Wang S et al (2017) Rice and cold stress: methods for its evaluation and summary of cold tolerance-related quantitative trait loci. Rice 7:24. https://doi.org/10.1186/s12284-014-0024-3

Zhao RR, Sheng JP, Lv SN et al (2011) Nitric oxide participates in the regulation of *LeCBF1* gene expression and improves cold tolerance in harvested tomato fruit. Postharvest Biol Technol 62:121–126

Zhao M, Liua W, Xiaa X et al (2014) Cold acclimation-induced freezing tolerance of *Medicago truncatula* seedlings is negatively regulated by ethylene. Physiol Plant 152:115–129

Zhu JK (2016) Abiotic stress signaling and responses in plants. Cell 167:313–324

Chapter 6
Cross Talk Between Cold Stress Response Signaling Pathway and Other Stress Response Pathways

V. C. Dilukshi Fernando

6.1 Introduction

Plants as sessile organisms are exposed to various environmental stresses including cold, heat, salinity, and drought. Plants respond to these environmental stimuli using different mechanisms. Some mechanisms have very specific signaling pathways that are unique to each stress, and some involve cross talk between multiple pathways (Huang et al. 2012). The promoter sequence elements in stress-responsive genes show evidence of response to different environmental stresses and thereby ability to perform functional adaptations to multiple stress conditions (reviewed in Nakashima et al. 2014). Abscisic acid (ABA) has been identified as the key plant stress hormone. However, these adaptations to various abiotic stress conditions can be either ABA dependent or ABA independent (Fig. 6.1) (Fernando and Schroeder 2016). Integration of cellular stress signal transduction pathways can better adapt plants to overcome the detrimental effects of stress conditions (Chen et al. 2008). In this mini review, we will briefly discuss the interactions between cold stress-responsive gene expression and signaling pathways of other environmental stresses. We will only focus on the major regulatory networks and only plant hormone ABA and cross talk between pathways.

6.2 Role of ABA in Stress Signal Transduction Pathways

ABA is a phytohormone with multiple roles. ABA acts as a signal transduction molecule, and ABA signaling pathways have been identified in plants using *Arabidopsis* as a model. Despite the function in plant growth and development,

V. C. D. Fernando (✉)
Department of Biological Sciences, University of Manitoba, Winnipeg, MB, Canada
e-mail: umfern34@myumanitoba.ca

S. H. Wani, V. Herath (eds.), *Cold Tolerance in Plants*,
https://doi.org/10.1007/978-3-030-01415-5_6

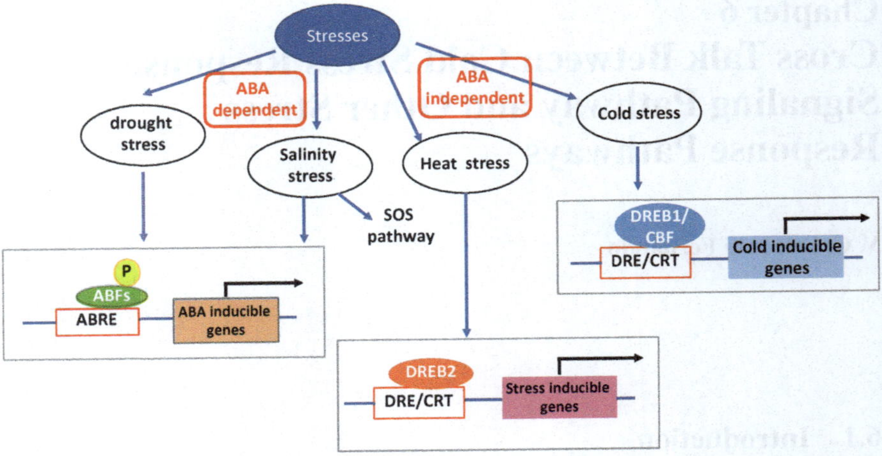

Fig. 6.1 Schematic diagram of transcription regulatory elements in different stress signaling pathways in plants

ABA is involved in plant water balance and osmotic stress tolerance in drought and salinity stress (Raghavendra et al. 2010). ABA acts as a messenger in salinity and drought stress to regulate the expression of downstream components of the stress signal transduction pathway. However not all stress signal transduction pathways occur via ABA (Shinozaki et al. 2003).

Cold acclimation is a gradual process where plants adapt to low temperatures below a threshold. ABA plays an important role in cold acclimation by promoting leaf senescence. At the same time, ABA integrates other hormonal pathways and reactive oxygen species (ROS) signaling to induce cold tolerance (Xue-Xuan et al. 2010).

6.2.1 ABA-Dependent Pathways

As indicated by Fig. 6.1, mainly osmotic stress induced by salt and drought triggers adaptive responses mediated by ABA to minimize or avoid the stress condition. High *ABA* levels in cells promote production of proteins like LATE EMBRYOGENESIS ABUNDANT (LEA) proteins in seeds as osmoprotectants, induce stomatal closure to minimize water loss, and reduce seed germination (Fernando and Schroeder 2016).

ABA-dependent gene expression requires ABSCISIC ACID BINDING FACTORS (ABFs). ABFs have different roles in multiple stress-responsive pathways. *ABF1* is induced by cold stress, while *ABF2* and *ABF3* are induced by salt stress. *ABF4* is known to be induced by salt, drought, and cold stress (Yoshida et al. 2010).

6.2.2 *ABA-Independent Pathways*

Cold stress signaling can be either ABA dependent or independent. C-repeat binding factor (CBF)/dehydration-responsive element binding (DREB) factor family of transcription factors has been identified to induce expression of cold stress-regulated (*COR*) genes in an ABA-independent manner (Ishitani 1997). In addition, NAC transcription factors also involve ABA-independent signal transduction in various abiotic and biotic stress conditions (Nuruzzaman et al. 2013).

6.2.3 *Transcription Factor Families in Stress Responses*

Transcription factor (TF) families that are known to play a role in cold stress responses are also induced by other stresses indicating cross talk between stress-responsive gene expression due to various other stress conditions (Kidokoro et al. 2015). Transcription factors play a major role in integrating hormone signaling pathways, and TFs work in concert or individually in these pathways (Seki et al. 2002). The major transcription factor families involved in stress signaling are shown in Table 6.1.

6.3 Cross Talk Between Stress Signaling Pathways

There is an overlap between stress-inducible genes and various environmental stresses. Salinity-induced genes are also drought inducible, and drought-inducible genes are also induced by cold stress (about 10%). In *Arabidopsis*, 277 genes were found to be drought inducible, 194 salinity inducible, and 53 cold inducible by cDNA microarray analysis. Thus, 53% of the genes were induced more than five-fold by drought stress suggesting water deficit condition is the major stress affecting plant growth. Out of the total of 524 genes upregulated by stress conditions, 22 genes (about 4%) were found to be induced by all three stresses (Seki et al. 2002). *RESPONSIVE TO DEHYDRATION29A (RD29A)/COLD REGULATED 78*

Table 6.1 Major transcription factor families in abiotic stress signaling (Nakashima et al. 2014)

Transcription factor	Binding domain	Stress signaling pathway
1. ABFs	ABSISIC ACID RESPONSIVE ELEMENT (ABRE)	Drought, salt, cold
2. DREB1/CBF	DEHYDRATION RESPONSIVE ELEMET (DRE)/C-REPEAT (CRT)	Drought, cold
3. DREB2	DRE/CRT	Dehydration, heat
4. NAC	NAC recognition sequence (NACRS)	Drought, high salt

(COR78)/LOW TEMPERATURE INDUCED (LTI78) is a stress-responsive gene that has been identified to be induced by all three stresses.

These genes induce accumulation of LEA proteins, heat-shock proteins (HSP), and KIN (cold-inducible) proteins that neutralizes ice nucleators, osmoprotectant proteins, carbohydrate metabolism-related proteins, water channel proteins, sugar transporters, senescence-related gene products, protease inhibitors, lipid transfer proteins, etc. (Seki et al. 2002).

The reason for the overlap of stress signaling pathways could be due to induction of osmotic stress by all three stresses: salt, drought, and cold. At the same time, each stress condition can lead to a unique signaling pathway by adapting plants to respond appropriately to each type of stress.

6.3.1 Cold Stress and Drought/Osmotic Stress

Freezing induces osmotic stress due to ice nuclear formation. During cold acclimation dehydration-responsive gene products are produced to overcome water-deficient conditions (Shinozaki et al. 2003).

Drought stress-responsive gene expression is highly ABA regulated, and the promoter region analysis of these genes shows a common element, ABA-responsive binding factors (AREB). Transcription factors known as ABFs are known to bind to these elements in drought-responsive genes (Choi et al. 2000). On the other hand, cold stress signal transduction occurs in an ABA-independent manner via the C-REPEAT/DEHYDRATION-RESPONSIVE ELEMENT BINDING FACTORS (CBFs/DREBs) (Shinozaki et al. 2003). Only 5% (30 genes) of stress-inducible genes were upregulated by both drought and cold (Seki et al. 2002). The reason could be the difference in stress signaling pathways and its dependence on ABA. Relatively ABA seem to play a minor role in cold stress tolerance.

DREB2 is an AP2-type TF which transactivates DRE/CRT in stress-responsive genes and thereby acts as a coupling element in drought/osmotic signaling (reviewed by Chinnusamy et al. 2007). This shows that osmotic signaling also occurs via ABA-independent pathways.

CBF4 was identified to be not responsive to cold stress but drought stress. CBF4 is a close homolog of CBF1/2/3 and similar to other CBFs. CBF4 also induces expression of COR genes. Expression of CBF4 is ABA dependent, whereas it is not so in the other CBFs. Thus CBF4 probably activates CRT/DRE via ABA which is the ABA-dependent regulation of cold stress signaling (Haake 2002).

6.3.2 Cold Stress and Salinity Stress

Like drought stress, salt also induces osmotic stress and in addition ionic stress which results in an imbalance in cellular homeostasis. Genetic and molecular analysis indicates dehydration-protectant proteins are produced in response to salt stress induced by an ABA-dependent pathway (Fig. 6.1) and therefore involve ABFs (Fernando and Schroeder 2016). In addition, salt stress signaling occurs via an ABA-independent salt overly sensitive (SOS) pathway where calcium acts as a secondary messenger and is activated to restore cellular homeostasis (Chinnusamy et al. 2007).

Seki et al. (2002) showed that in *Arabidopsis* 24 genes (out of 524) were induced by both salt and cold stress. There were more genes (75%) induced by drought and salt than salt and cold indicating a stronger relationship between drought and salt stress. They also noted 194 genes that were highly induced by salt, which results in accumulation of LEA proteins, water channel proteins that regulate cellular osmotic pressure, and potassium transporters which play an important role in regulating Na^+ and K^+ uptake during salt stress (Seki et al. 2002).

6.3.3 Cold Stress and Heat Stress

Temperature stress (both heat and cold) induces a series of cellular changes including membrane fluidity changes, calcium signaling, protein degradation, and production of ROS (Chinnusamy et al. 2004). Heat stress produces Heat Shock Proteins (HSPs) which are chaperones that prevent protein degradation. Heat stress also induces mitogen-activated protein kinases (MAPKs). MAPK cascades are important regulators of multiple abiotic stress responses and therefore are a key component of cross talk between different stress signal transduction pathways (Huang et al. 2012).

Both DREB1 and DREB2 bind to DRE/CRT *cis*-elements on stress-responsive genes. However, DREB1 has a more important role in cold stress while DREB2 in heat stress. Studies show overexpression of *DREB2* results in production of HSPs that regulates thermotolerance in plants. DREB2 not only functions in thermotolerance but also in dehydration tolerance. Downstream gene products of *DREB2* and *DREB1* indicate they have similar functions except for carbohydrate metabolism (Sakuma et al. 2006). Although studies were mainly on *Arabidopsis*, the function of DREB2s has also been studied in other crop species such as rice and soybean (Nakashima et al. 2014).

6.4 Implications of Interactions Between Stress Signaling Pathways

Stress signaling cascades regulated by calcium, calcium-regulated proteins, and MAPKs play an important role together with TFs to help plants adapt to numerous stress conditions (Huang et al. 2012). As indicated in Chinnusamy et al. (2007), MAPKs are activated by multiple stresses, and therefore they act as a hub for different stress signaling pathways. Stress-responsive genes are induced by more than one type of stress, indicating cross talk between multiple pathways. Interaction between pathways makes a more efficient stress signaling network within plants that make them better adapt to various environmental stresses. Understanding cross talk between pathways also enables improving plant cold stress tolerance in a biotechnological perspective.

References

Chen H, Zhang J, Neff MM, Hong S-W, Zhang H, Deng X-W, Xiong L (2008) Integration of light and abscisic acid signaling during seed germination and early seedling development. Proc Natl Acad Sci 105(11):4495–4500. https://doi.org/10.1073/PNAS.0710778105

Chinnusamy V, Schumaker K, Zhu JK (2004) Molecular genetic perspectives on cross-talk and specificity in abiotic stress signalling in plants. J Exp Bot 55(395):225–236. https://doi.org/10.1093/jxb/erh005

Chinnusamy V, Zhu J, Zhu JK (2007) Cold stress regulation of gene expression in plants. Trends Plant Sci 12(10):444–451. https://doi.org/10.1016/j.tplants.2007.07.002

Choi H, Hong J, Ha J, Kang J, Kim SY (2000) ABFs, a family of ABA-responsive element binding factors. J Biol Chem 275:1723–1730

Fernando VCD, Schroeder DF (2016) Role of ABA in Arabidopsis salt, drought, and desiccation tolerance. In: Abiotic and biotic stress in plants – recent advances and future perspectives, pp 507–524. Intechopen, Rijeka Croatia https://doi.org/10.5772/61957

Haake V (2002) Transcription factor CBF4 is a regulator of drought adaptation in Arabidopsis. Plant Physiol 130(2):639–648. https://doi.org/10.1104/pp.006478

Huang GT, Ma SL, Bai LP, Zhang L, Ma H, Jia P, Liu J, Zhong M, Guo ZF (2012) Signal transduction during cold, salt, and drought stresses in plants. Mol Biol Rep 39(2):969–987. https://doi.org/10.1007/s11033-011-0823-1

Ishitani M (1997) Genetic analysis of osmotic and cold stress signal transduction in Arabidopsis: interactions and convergence of abscisic acid-dependent and abscisic acid-independent pathways. Plant Cell 9(11):1935–1949. https://doi.org/10.1105/tpc.9.11.1935

Kidokoro S, Watanabe K, Ohori T, Moriwaki T, Maruyama K, Mizoi J, Myint Phyu Sin Htwe N et al (2015) Soybean DREB1/CBF-type transcription factors function in heat and drought as well as cold stress-responsive gene expression. Plant J 81(3):505–518. https://doi.org/10.1111/tpj.12746

Nakashima K, Yamaguchi-Shinozaki K, Shinozaki K (2014) The transcriptional regulatory network in the drought response and its crosstalk in abiotic stress responses including drought, cold, and heat. Front Plant Sci 5(May):1–7. https://doi.org/10.3389/fpls.2014.00170

Nuruzzaman M, Sharoni AM, Kikuchi S (2013) Roles of NAC transcription factors in the regulation of biotic and abiotic stress responses in plants. Front Microbiol 4:248. PMC. Web. 9 July 2018

Raghavendra AS, Gonugunta VK, Christmann A, Grill E (2010) ABA perception and signalling. Trends Plant Sci 15(7):395–401

Sakuma Y, Maruyama K, Qin F, Osakabe Y, Shinozaki K, Yamaguchi Shinozaki K (2006) Dual function of an Arabidopsis transcription factor DREB2A in water-stress responsive and heat-stress-responsive gene expression. Proc Natl Acad Sci U S A 103:18822–18827

Seki M, Narusaka M, Ishida J, Nanjo T, Fujita M, Oono Y, Kamiya A et al (2002) Monitoring the expression profiles of 7000 Arabidopsis genes under drought, cold and high-salinity stresses using a full-length CDNA microarray. Plant J 31(3):279–292. https://doi.org/10.1046/j.1365-313X.2002.01359.x

Shinozaki K, Yamaguchi-Shinozaki K, Seki M (2003) Regulatory network of gene expression in the drought and cold stress responses. Curr Opin Plant Biol 6(5):410–417. https://doi.org/10.1016/S1369-5266(03)00092-X

Xue-Xuan X, Hong-Bo S, Yuan-Yuan M, Gang X, Jun-Na S, Dong-Gang G, Cheng-Jiang R (2010) Biotechnological implications from Abscisic acid (ABA) roles in cold stress and leaf senescence as an important signal for improving plant sustainable survival under abiotic-stressed conditions. Crit Rev Biotechnol 30(3):222–230. https://doi.org/10.3109/07388551.2010.487186

Yoshida T, Fujita Y, Sayama H, Kidokoro S, Maruyama K, Mizoi J, Shinozaki K, Yamaguchi-Shinozaki K (2010) AREB1, AREB2, and ABF3 are master transcription factors that cooperatively regulate ABRE-dependent ABA signalling involved in drought stress tolerance and require ABA for full activation. Plant J 61:672–685

Chapter 7
Proteomic Responses to Cold Stress

Towseef Mohsin Bhat, Sana Choudhary, and Nirala Ramchiary

7.1 Introduction

Population growth rates globally have so outstripped the linear rate of increases in food production that the Food and Agriculture Organization of the United Nations (FAO) estimated that 70% more food must be produced over the next four decades in order to nourish adequately a human population projected to exceed 9 billion by the year 2050 (Rosenzweig and Parry 1994). Plant "genome," the total genetic constitution consisting of DNA both coding and noncoding, is under constant stress from endogenous as well as exogenous factors due to the generation of reactive oxygen species (ROS) in response to abiotic stress including the cold stress (Virdi et al. 2015). Cold stress is a vital environmental influence that confines the agricultural yield of plants in colder areas. Timperio et al. (2008) reported that low temperature exhibits a negative impact on plant productivity mostly because plant metabolism and physiology change due to cold stress. Plants outcompete the cold stress at molecular, cellular, physiological, and biochemical levels through expression of numerous genes, proteins, and secondary metabolites (Sanghera et al. 2011). Plants have developed a detailed mechanism which enables them to receive the external signals and to evade various adaptive responses with physiological alterations (Hashimoto and Komatsu 2007). Barrero-Gil and Salinas (2013) and Miura and Furumoto (2013) suggested that cold stress adaptation occurs by modification in gene expression and metabolism of plants through transforming in regulation of transcriptional posttranscriptional regulation as well as translational and posttranslational regulation of cold stress signaling during acclimatization. First and foremost

T. M. Bhat (✉) · N. Ramchiary
Laboratory of Translational and Evolutionary Genomics, School of Life Sciences
Jawaharlal Nehru University, New Delhi, India

S. Choudhary
Genetics Section, Department of Botany, Aligarh Muslim University, Aligarh, India

S. H. Wani, V. Herath (eds.), *Cold Tolerance in Plants*,
https://doi.org/10.1007/978-3-030-01415-5_7

we have to summarize criteria through which we can say that the plant is facing cold. The best definition of cold was quoted by Christian Korner (2016) which says, "I refer to "cold" as free atmospheric conditions that deviate substantially (negatively) from optimum temperatures for growth in most vascular plant taxa on the globe, causing severe growth limitations (periodically no or very little growth for temperature reasons, or the risk of tissue damage and survival)". As far as cold stress is concerned, plants can exhibit wide and broad range of responses, caused by cold stress, such as physiochemical response which occur together with gene expression fluctuations (Heidarvand and Amiri 2010), and because of these changes, plants become more resistant to cold climate. Crop productivity is indirectly and heavily affected by low temperatures leading to crop erosion. Different plants have different tolerance mechanisms to cope with the chilling (0–15 °C) and freezing (< 0 °C) temperatures. Plants growing in temperate climates tolerate chilling stress with inconsistent degrees of chilling tolerance during prevailing nonfreezing temperatures. The chilling resistance also known as chilling tolerance is always associated with physiological and biochemical changes and finally leads to marker changes in gene expression and differential protein expression for the synthesis of membrane lipids and proteins. A number of proteins have been observed through comparative proteomic analysis, which are newly synthesized, induced, or repressed in response to low temperature in different crop species such as in *Arabidopsis* (Fanucchi et al. 2012), wheat (Rinalducci et al. 2012), rice (Neilson et al. 2011), strawberry (Gu et al. 2013), and potato (Folgado et al. 2013). These proteins whose functions are directly involved in cold and freezing tolerance include dehydrins, late embryogenesis abundant (LEA) proteins, cold-regulated (COR) proteins, antifreezing proteins, and pathogenesis-related (PR) proteins (Gharechahi et al. 2016). Plants which grow in tropical and subtropical climate regimes are very much sensitive to chilling stress and low mechanism of cold acclimation. Understanding cold acclimation (CA) is important for concurrently increasing autumn yield and winter survival in plants.

Cold stress limits plant growth, distribution, and development. Understanding how plants transduce and respond to cold signals has long been a topic of interest for many molecular biologists and physiologists. Traditional genetic and molecular analyses have identified C-repeat/DREB-binding factors (CBFs) as key transcription factors that function in cold acclimation. Recent studies revealed the involvement of pivotal protein kinases and transcription factors in CBF-dependent signaling, expanding our knowledge of cold signal transduction from perception to downstream gene expression events. Knowledge of the mechanism underlying the ability of plants to survive freezing temperatures will facilitate the development of crop plants with increased freezing tolerance (Shi et al. 2018).

Proteomics is an important molecular mechanism for determining the complete proteome of the cell, organelle, organ, or tissue under given physiological conditions (Sobhanian et al. 2010). It was suggested that different proteomic approaches have emerged as important phenomenon to identifying genes and pathway which are related to stress response and tolerance (Abdalla and Rafudeen 2012; Gao et al. 2011; Zhang et al. 2010). In the present day, the two-dimensional electrophoresis (2DE) and two-dimensional difference gel electrophoresis (2D-DIGE) followed by mass spectrometry (MS) have been mostly used to investigate changes in chilling and freezing stress-related proteins in various crops such as *Arabidopsis thaliana*, soybean, barley, rice, and wheat (Nakaminami et al. 2014; Cheng et al. 2010; Hlaváčková et al. 2013; Lee et al. 2009; Kosová et al. 2013; Gharechahi et al. 2014).

This chapter throws light in understanding of proteomic response under cold stress and protein extraction difficulties, differential expression of proteins, iTRAQ proteomics, plant proteomic approaches and challenges, and recent innovative technologies used in quantitative plant proteomics for cold stress. In this chapter proteomic tools and the challenges faced with cold stress by plants are discussed. Subsequently, the function of cold responsive proteins is reviewed in the context of cold stress tolerance in plants. Additionally, this chapter will critically evaluate the current literature on the plant proteomics for understanding plant stress responses in detail in addition to the basic concept, principles, and procedure outlines of these approaches and will also suggest important future perspectives.

7.2 Impact of Cold Stress on Plants

Plant cold stress refers to a damage that is caused by a temperature drop to below 15 °C but above the freezing point, and the most common site cold injury is implicated in plasma membrane. The consequence of this change may lead to cell leakage or disruption. Changes in cell permeability are often invoked as a cause of the loss of cell turgor. Many cold stress symptoms are common to other stresses such as drought stress, phytotoxicity stress, heavy metal stress, heat stress, and light stress. Primarily, it involves rapid wilting followed by water-soaked patches which develop into sunken pits that reflect cell tissue collapse. Formation of crystalline deposits in root, epidermal, vascular, and mesophyll cells of leaves leading to tonoplast disruption accelerated rate of senescence (natural death) but with otherwise normal appearance and slow growth. Cold injury causes several metabolic and physiological dysfunctions to the plant including disruption of the conversion of starch to sugars (amylolytic activity),decreased carbon dioxide exchange, reduction in net photosynthesis, destruction and degradation of chlorophyll, decrease in respiratory activity, decrease in fatty acid synthesis which lead to starvation in some plants, and finally death. Low-temperature stress or freezing stress causing injury in plants can be from two sources: (a) freezing of soil water and (b) freezing of the fluids within the plant. The soil water that is available to the plants is found in the porous regions between soil particles; it freezes at about −2 °C depriving the plant of its source of water and causes disruption of cells and tissues of the plant body. Pearce and Humphrey (2001) observed that plants freeze when they cannot avoid nucleation (water molecules come together to form a stable ice nucleus) and cannot prevent the growth of ice. Freezing-point depression, caused by the presence of solutes (1–2 °C: Levitt 1980; Franks 1985) and by super cooling, is often too slight to prevent freezing in moist temperate and colder climates. Consequently, plants of contrasting types nucleate at mild temperatures.

In the field, peach freezes between −0.6 and −2.6 8 °C, and other overwintering temperate woody species freeze between −1.2 and −2.1 8 °C (Ashworth 1992): grasses can freeze between −1.5 and −2.5 °C (Pearce and Fuller 2001). Laboratory tests often give lower nucleation temperatures than those reported under natural conditions and so can be unreliable indicators of behavior in the field (Flinn and Ashworth 1994) (Fig. 7.1).

There are various morphophysiological, biochemical, and molecular mechanisms which are involved during the response of plants to low-temperature or chilling stress. With ongoing improvement in the systems biology and proteomics era, the fascination in deciphering protein overexpression and its influence on the cellular microenvironment of crop plants forms an interesting part of plant proteomics (Taylor et al. 2005; Bae et al. 2003). iTRAQ-based proteomics, the cutting-edge molecular technique of the present time, offers various advantages over the genome-based technologies as it precisely dispenses with the functional molecules as well as posttranslational modifications (PTMs) rather than genetic code or mRNA abundance (Walsh et al. 2010). Higgins and Hames (1999) and Kwon et al. (2002) have reported protein expression mechanisms earlier, and the number is increasing every year. Protein function is determined by its folding, and the modification of amino acids and their side chains contributes remarkably to the structural and functional diversity of the proteins involved in chilling stress.

7.2.1 Protein Extraction from Notorious Plant Tissues

Protein extraction from plant tissues is a very difficult job in terms of protein concentration and purity. There are various protocols given by various scientists from time to time for the extraction of relevant proteins from tissues. Here, we only give a troubleshooting guide for protein extraction. The best protein protocol which works for most of the plant species is the protocol "Extraction of Total Proteins from Rice Plant by Da-Gin Lin and Chang-Sheng Wang" (http://www.bio-protocol.org/e1277); this type of protocol shows materials and reagents used for the extraction process. Table 7.1 shows problems and solutions during protein extraction processes.

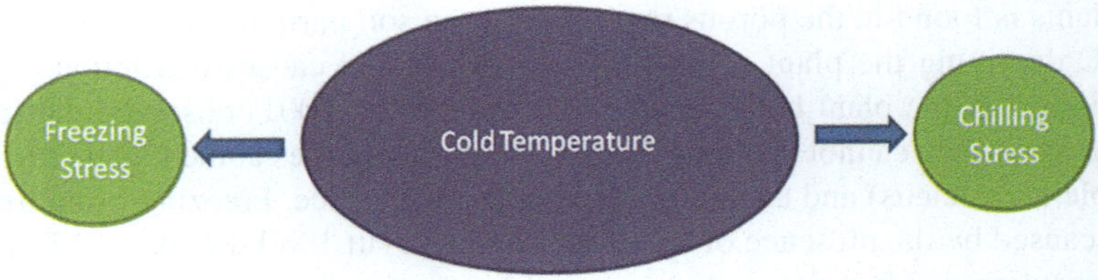

Fig. 7.1 Proteomics as a promising tool in the study of plant response to cold stress

Table 7.1 Troubleshooting guide during protein extraction

Problem	Cause	Solution
Low protein yield	Insufficient tissue disruption	Grind tissue in liquid nitrogen to a fine powder. Do not allow the tissue to thaw. Fibrous tissue such as maize leaf, all stems, and some roots may require additional grinding. Addition to grinding aids, such as clean high-quality sand, may be used as a last resort
	Old or dry starting tissue	Use young healthy plant tissue for protein extractions. Avoid old or dry tissue
	Low protein content in starting tissue	Some plant tissue does not contain large quantities of protein. Adjustment may be made by decreasing the volume of the extraction reagent or by increasing the amount of the tissue. One may do additional extractions and combine them; however, concentration step may be required with this approach
Poor quality protein recovered	High phenolics or tannin concentration	Additional methanol and/or acetone washes may be required for large quantities of the tissue such as a pine needle or cotton leaf. If there is any green color remaining in the methanol or acetone washes, repeat washes until colorless
	Protein degradation	Make certain the protease inhibitor cocktail for plant extracts is included in the appropriate solutions. Do not allow the tissue to thaw while grinding or before extraction with methanol
	Formation of protein complexes	Many plant metabolites will complex with proteins. If these complexes form, they are nearly impossible to break apart. These complexes will interfere with many downstream applications
	Large molecular weight contaminants	Make certain the final protein extract is centrifuged for 30 min at 16,000 × g to pellet any contaminants. Be careful not to disrupt the pellet in any of the steps of the procedure

7.2.2 Differential Expression Proteomics

Through comparison and composition of proteins, differential proteomics was born. Biologists working on stress proteomics compare the protein expression of non-stressed plants acting as controls with stressed ones (subjected to varied degrees of stress). Most studies in differential proteomics compare the total proteins of contrasting plant genotypes and search for potential targets for raising crops under various hostile environmental conditions including chilling and freezing stress. Another important study is to compare proteomes of two different species of plants with varying degrees of tolerance for a particular stress factor. For these types of studies where two contrasting proteomes are differentiated, two-dimensional electrophoresis (2DE) followed by protein identification using MALDI TOF-S analysis is applied. MS-based differential proteomics involved in cold response have been identified by various workers.

Tian et al. (2011) studied proteomics of *Arabidopsis* by two-dimensional difference gel electrophoresis (2D DIGE) analysis of subcellular fractions of *Arabidopsis thaliana* proteome coupled with spot identification by tandem mass spectrometry and revealed posttranslational mechanisms involved for cold-induced metabolic changes. They observed that cold-induced fast metabolic changes are involved for the protection of cold-induced damage before temperatures decrease to freezing temperature. The proteins identified include four enzymes involved in starch degradation, three HSP100 proteins, several proteins of Krebs cycle, and sucrose metabolism. They also observed that upon cold treatment, the disproportionating enzyme 2 (DPE2), a cytosolic transglucosidase metabolizing maltose to glucose, increased rapidly. Cook et al. (2004) determined that CBF cold response pathway is involved in configuring the low-temperature metabolome in *Arabidopsis*. Cui et al. (2005) studied the proteome analysis of cold responses in rice (*Oryza sativa*) seedlings and isolated 1700 proteins based on differential proteomics. They found cold-responsive proteins, besides two proteins of unknown function which included four factors of protein biosynthesis, four molecular chaperones, two proteases, eight enzymes involved in biosynthesis of cell wall components, seven anti-oxidative/detoxifying enzymes, and proteins linked to energy pathway, as well as a protein involved in signal transduction. They found that the functional proteomes illuminate the facts, at least in plant cell, that protein quality control mediated by chaperones and proteases and enhancement of cell wall components play important roles in tolerance to cold stress. Cold-responsive protein PEPC2 associated with the TCA cycle was observed by Xuan et al. (2013) in stolon of a C4 perennial grass species and proposed a cold stress-responsive protein network composed of several different functional components. These proteins exhibited a balance between reactive oxygen species (ROS) production and scavenging, accelerated protein biosynthesis and proteolysis, reduced protein folding, enhanced photosynthesis, abundant energy supply, and enhanced biosynthesis of carbohydrates and nucleotides. PEPC catalyzes the reaction of β-carboxylation of phosphoenolpyruvate to yield oxaloacetate and phosphate—a reaction involved in several metabolic pathways (Sanchez et al. 2006).

7.3 iTRAQ Proteomics in Cold Stress

Recently, iTRAQ (isobaric tags for relative and absolute quantitation)-based proteomics have revolutionized the protein chemistry field. It is an isobaric labeling method used in quantitative proteomics in tandem mass spectrometry to determine the amount of proteins from different sources in a single experiment. It uses stable isotope-labeled molecules that can be covalent bonded to the N-terminus and side chain amines of proteins. The balance groups present in each of the iTRAQ reagents function to make the labeled peptides from each sample isobaric, and the quantification is facilitated through analysis of reporter groups that are generated upon fragmentation in the mass spectrometer. There are currently two mainly used reagents, 4-plex and 8-plex, which can be used to label all peptides from different samples/treatments. These samples are then pooled and usually fractionated by nano-liquid chromatography and analyzed by tandem mass spectrometry (MS/MS).

Cold stress proteomics is aimed at comparing several proteomes dominated by 2DE followed by protein identification via MS analysis, although the sole use of MS techniques is not only for stress protein identification but also for stress protein quantitation.

Isobaric tags for relative and absolute quantitation (iTRAQ) coupled with liquid chromatography-quadrupole mass spectrometry (LC-MS/MS) describes a recently developed technique which provides a fast proteomic analytical method for the identification and quantification of expressed proteins with a high degree of efficiency and accuracy (Evans et al. 2012) and is currently being widely used for the quantitative comparative analysis of plant proteomes (Owiti et al. 2011; Zheng et al. 2014).

Zhang et al. (2016) studied "quantitative proteomics" analysis for investigating novel insights into cold stress responses in *Petunia* seedlings using iTRAQ and highlighted the role of antioxidation mechanisms and epigenetic factors in the regulation of cold stress responses. Bae et al. (2003) analyzed *Arabidopsis* nuclear proteome and its response to cold stress. They analyzed nuclear proteins using two-dimensional (2D) gel electrophoresis and matrix-assisted laser desorption/ionization time of flight mass spectrometry (MALDI-TOF MS). Approximately 500–700 spots were detected in reference 2D gels of nuclear proteins. Proteomic analyses led to the identification of 184 spots corresponding to 158 different proteins implicated in a variety of cellular functions. Liu et al. (2017) also studied iTRAQ-based quantitative proteome metabolic changes of *Flammulina velutipes* mycelia in response to cold stress and identified and quantified a total of 1198 proteins in three different mycelium growth stages of *F. velutipes*, demonstrating the applicability of the iTRAQ approach to multiplexed proteomic profiling. Rinalducci et al. (2011) analyzed proteome of a spring wheat cultivar in response to prolonged cold stress and represented cold as one of the major abiotic factors influencing plant growth and development worldwide. They analyzed the long-term responsiveness of Iranian spring wheat (cv. Kohdasht) to cold from a proteomic point of view, in order to unravel the molecular mechanisms helping a cold-sensitive cultivar to survive exposure to suboptimal temperatures. Quantitative analyses on protein alterations occurring upon low-temperature exposure showed a reinforcement in ascorbate recycling (dehydroascorbate reductase, ascorbate peroxidase) and protein processing (proteasome subunit, cysteine proteinase), as well as the accumulation of the enzyme devoted to tetrapyrrole synthesis (glutamate semialdehyde aminomutase). In contrast, among proteins downregulated after cold stress, we could identify some key Krebs cycle enzymes (isocitrate dehydrogenase, malate dehydrogenase), together with many photosynthesis-related proteins (oxygen evolving complex proteins, ATP synthase subunits, ferredoxin NADPH oxidoreductase, and some Calvin cycle enzymes). Physiological and biochemical parameters (such as shoot apex dissection, chlorophyll, proline, and sugar content determination) sustained proteomic findings allowing the present research to contribute to the current knowledge on these responses, which may be crucial to stress adaptation under field conditions.

7.4 Plant Proteomic Approaches and Challenges for Cold Tolerance

Although productivity of world ecosystems is probably limited more by water than any other environmental factor, low temperature is probably most limiting to plant distribution (Burke et al. 1976). To grow even in subtropical regions subject occasionally to near-freezing temperatures, plants must be capable of some acclimation to low temperatures; plants that grow in polar regions must tolerate extremely low temperatures and only important hazard in most agricultural regions of the world. Low temperatures affect mostly plants like citrus, cotton, maize, rice, sorghum, sugarcane, and sweet potato. Rice plants exposed to a low temperature of about 16 °C may widely affect its pollen fertility, and if exposed to low temperature, pollen mother cell division will not produce crop. It has been estimated that the world rice production would decrease by 40% if world mean temperature dropped only 0.5–10 °C. Proteomic studies of membranes and lipid model of membranes suggest that the membrane normally existed in a liquid crystalline condition; in this state the enzymes have their optimal activity, and its permeability is thus under control. Below the freezing temperatures, the membrane exists in a solid gel state; this change in state would bring about a contraction resulting in cracks or fissures that increase permeability. This would lead to the irregular solute balances from the mitochondria and other organelles and also lead to enzyme denaturation; optical activity of most of the membrane proteins is also lost by low-temperature stress. Low-temperature stress is often defined as decrease in the temperature level which is sufficient to cause irreversible damage to plant growth and development. Low- or high-temperature stress causes morphological anatomical and physio-biochemical changes in plants, which adversely affect plant growth and development and may lead to a drastic reduction in economic yield (Hasanuzzaman et al. 2013). The adverse effects of cold stress can be mitigated by developing cold-tolerant plants using various physiological and genetic approaches. For this purpose a thorough understanding of mechanisms of cold tolerance, physiological responses of plants to low temperature, and possible strategies for improving crop productivity under stress condition is imperative. Nowadays, special attention has been provided in linking genomic and transcriptomic profiles to specialized expression, function, and protein functional networks in various plant model species which will act as references for other plants. While studying complexity and dynamicity of plant protein expression, it is hard to choose or focus on reliable proteomic approaches targeting the identification of proteins and their modification that may contribute to crop improvement. Nowadays, various efficient and high-throughput technologies have been contributing in the studying of quantitative proteomics and understanding plant growth-related stress markers.

All cellular functions in living organisms are carried out by proteins. It is well known that most proteins are not active alone but interact with other proteins to form complexes. The interaction of proteins and the formation of protein complexes have been studied for many years with a range of in vitro and in vivo methods. Most of these methods only allowed the identification of a limited number of proteins. These include yeast-two-hybrid interaction studies, co-immunoprecipitations, and bi-fluorescence complementation assays. Commonly used 2D gels combined with mass spectrometric analyses can also not deliver more than a few hundred proteins within one experiment. Moreover, none of the methods mentioned above allow reliable quantification of protein abundances. Native gel electrophoresis and gel filtration to fractionate protein complexes from plant extracts are somehow a reliable and applicable method to study stress-related proteomics. Combination of classical methods with high-throughput mass spectrometric analysis, which allow the detection and quantification of several thousand proteins in one experiment, is way forward to study the complexity of protein expression during low-temperature stress. In the first phase of this study, a protein scientist has to establish a map of protein complexes in *Arabidopsis*. The confirmation of previously published protein-protein interactions will serve as a backbone to which predictions of new interactions will be added. Experiments to confirm predicted components of protein complexes will include tap-tag immunoprecipitations, and functional relationships will be unraveled using classic genetic and biochemical methods. In the second phase, the established map will be used to quantitatively study changes in protein complex abundances and complex composition in different environmental conditions and genetic backgrounds. These experiments will enable us to draw conclusions about the role of protein complex formation in the regulation of metabolic pathways in various plants.

7.5 Recent Innovative Technologies Used in Quantitative Plant Proteomics

Driven by innovations in MS-based technologies and rapid development of quantitative methods, proteomics has emerged as a complementary technique to other approaches such as transcriptomics and metabolomics in the post-genomic era (Wienkoop et al. 2010). Proteomic analysis is achieved through (1) separation and identification of proteins based on 2DE or coupled gel-free shotgun liquid chromatography-tandem mass spectrometry (LC-MS/MS) platforms; (2) elucidation of protein functions and protein functional networks in plant metabolic and signaling pathways through the analysis of protein mapping, characterization of PTMs, and protein-protein interactions; and (3) bioinformatics strategies and the use of databases for both model and non-model plant species (Holman et al. 2013). Recently, the application of gel-free protein separation approaches and "second-generation" proteomic techniques such as multidimensional protein identification technology (MudPIT), quantitative proteomic approaches including isotope-coded affinity tags (ICATs), targeted mass tags (TMTs), and isobaric tags for relative and absolute quantitation (iTRAQ) has been widely used in descriptive and comparative proteomic studies of plant development and metabolic strategies in abiotic stress adaptation. Advances in liquid chromatography-based separation and label-free quantitative proteomic analysis of a large number of proteins derived from complex plant samples have recently been discussed (Matros et al. 2011). Although modern gel-free quantitative proteomic approaches, label-based and label-free, are considered to be more advanced and can provide more information on comparative changes in protein expression than one and 2DE gel-based methods which have limitations. One obvious limitation in terms of global proteome coverage is the fact that they are designed for less hydrophobic, more aqueous buffer-soluble sub-proteomes, whereas the buffers and detergents used in gel-based protein separation techniques can be quite powerful and efficient in solubilization of more hydrophobic protein groups. In combination with one and 2DE gels was the new realization that helped to solve one of the main drawbacks of gel-based separation and quantitative analysis of proteins related to co-focusing/co-localization of several proteins and their modified forms in one spot or band on the gel. However, until recently, large-scale proteomic studies with unkown organisms were difficult to realize due to the lack of genomic information available to facilitate protein identification. The publication of the first draft of completely sequenced wheat genome (International Wheat Genome Sequencing Consortium [IWGSC] 2014) is inspirational to plant proteomic researchers, although the annotation of complete wheat genome assembly will remain a challenging task. Until recently proteomic studies have been using alternative available bioinformatics resources that included wheat EST-based databases (Bykova et al. 2011) or available closely related *Brachypodium distachyon* model plant genome, or D-genome progenitor *Aegilops tauschii*, as well as a composite database of available cereals sorghum, maize, and rice (Pascovici et al. 2013), a translated database of the low copy number genome assemblies of *T. aestivum*, and proteins from monocot family Poaceae (Kang et al. 2016). Pascovici et al. (2013) have evaluated the most effective pipeline for large-scale shotgun quantitative experiments using bread wheat (*T. aestivum*), iTRAQ multirun quantitative approach, and the available resources for bioinformatics data analysis and downstream functional interpretation. The study emphasized many challenges related to the repetitiveness/redundancy/polymorphism of bread wheat genome and therefore extremely large size of the corresponding EST-based database that could not be readily manipulated by the available bioinformatics tools and the stochastic aspect of protein grouping across multiple runs. The use of smaller databases was demonstrated as alternative pragmatic approaches to reliably identify proteins and proceed with functional annotations. More recently, targeted MS-based quantitative approaches such as multiplexed selective reaction monitoring (SRM) have proven to be powerful for identification of specific proteins with causative functions in agronomically important traits. Attributable to outstanding sensitivity of this methodology to selective quantitation of low abundance protein components in complex mixtures, SRM technique is seen by researchers as an alternative to antibody-based immune detection assays (Picotti et al. 2013). This SRM approach is based on highly specific detection and quantitation of proteotypic couples comprised of target precursor and corresponding fragment ions, and until recently it was exclusively based on triple quadrupole MS platforms due to the necessity of two stages of mass filters. The advantages and limitations of this application have been discussed and demonstrated experimental elsewhere. The new generation of Orbitrap technology instruments such as Q Exactive hybrid quadrupole-Orbitrap provides an efficient and user-friendly alternative for further application of SRM method in quantitative assay development with high potential for large-scale targeted proteomic experiments. The purpose of SRM methodology in plant proteomics is biomarker validation in crops, which follows the discovery phase with more explorative qualitative and quantitative comparative proteomic studies aimed at finding potential candidates important in stress responses. This highly selective and sensitive quantitative approach can be powerful not only for biomarker validation but also for the development of new stress tolerance assessment methods, which will facilitate the identification of genotypes with improved resistance and ultimately discovery of gene targets for marker-assisted breeding. Another approach has recently been developed for label-free shotgun proteomics based on data-independent (MSE) acquisition protocols that has a potential to identify peptides from complex samples in a rapid, consistent, and sensitive way and overs a higher dynamic range for peptide quantification (Buts et al. 2014). In this approach, ultra-performance liquid chromatography is used, which is coupled to an LC-MS/MS run where an alternating energy level allows to obtain accurate precursor masses at low energy and to take fragmentation spectra of all parent masses at high collision energy in one analytical run. However, this approach is rather at the early stages of development and has many challenges related to the difficulties in interpretation of very complex composite fragmentation spectra resulting in poor protein identification rate. At present, parallel data-dependent runs are needed in order to acquire all necessary information, build the databases of individual peptide fragmentation spectra, and link them to the MSE (Buts et al. 2014). This approach was successfully used in quantitative analysis of important allergenic proteins in wheat grain extracts, in identification of gliadins and glutenins in wheat grain and quantitation of proteins associated with celiac disease and baker's asthma (Uvackova et al. 2013a, b). An alternative strategy that combines high-specificity data-independent acquisition method with a novel targeted data extraction approach to mine the resulting fragment ion data sets was recently demonstrated (Gillet et al. 2012). This method, termed SWATH MS, is based on sequential time- and mass-segmented acquisition, which generates fragment ion spectra of all precursors in two user-defined dimensions, retention time and m/z space, resulting in complex fragment ion maps. The interpretation of highly specific multiplexed data sets required the development of fundamentally different data analysis strategies, which uses previously acquired information contained in spectral libraries to mine the fragment ion maps for targeted extraction and quantitation of specific peptides of interest. The accuracy and consistency of SWATH MS was demonstrated to be comparable to SRM approach (Gillet et al. 2012).

One of the important advantages of the former, alleviating most constrains of present proteomic methods, is the iterative retrospective re-mining of the acquired data sets for targeted extraction. This approach offers unprecedented possibility for the qualitative and quantitative profiling not only in proteomics but also in metabolomics and lipidomics. One of the main bottlenecks for proteomic development is the lack of robust bioinformatics tools with novel algorithmic solutions to processing of MS data, which are lagging behind the substantial advances occurring in instrumentation and protocols (Cappadona et al. 2012). It remains to be seen how this breakthrough technology will evolve into a powerful tool utilized throughout the plant.

7.6 Conclusion

The study of proteomic responses to cold stress and temperature interactions is of great relevance with respect to the current global climatic changes. The current chapter covers different topics related to proteomics and cold signal transduction mechanisms under which plant develops tolerance against cold stress. A critical factor limiting the productivity and distribution of many plant and crop species is the cold stress. Low temperature is initially recognized by plasma membrane either due to modifications in membrane fluidity or with the help of various sensory proteins like Ca2+-permeable channels, histidine kinases, receptor kinases, and phospholipases that finally lead to cytosolic Ca2+ influx. These proteins identified recently using various proteomic approaches. The signals of cold stress are transduced to the nucleus in order to switch on transcriptional cascades. Thus, understanding the mechanism of different protein signaling networks; role of various transcriptional factors, genes, and their products in regulating cold stress response; and cross talk among various signaling components should remain an area of intense research activity in the near future. As every abiotic stress involves multigenic traits, therefore, a multi-interdisciplinary approach involving physiological and biochemical analyses aided by proteomic- and genomic-based platforms should be followed in order to develop novel methods of analysis just to acquire better knowledge of gene expression. The information gathered through these studies finds great application in genetic engineering to develop transgenic plants that possess better tolerance to various abiotic stress conditions without showing any growth and yield penalty. Future research for acquiring the cold stress tolerance would greatly be based on utilization of high-throughput techniques developed in recent times, in combination with conventional genetic and breeding protocols and techniques.

References

Abdalla KO, Rafudeen MS (2012) Analysis of the nuclear proteome of the resurrection plant Xerophyta viscosa in response to dehydration stress using iTRAQ with 2DLC and tandem mass spectrometry. J Proteome 75(8):2361–2374

Ashworth EN (1992) Formation and spread of ice in plant tissues. Hort Rev 13:215–255

Bae MS, Cho EJ, Choi EY, Park OK (2003) Analysis of the Arabidopsis nuclear proteome and its response to cold stress. Plant J 36(5):652–663

Barrero-Gil J, Salinas J (2013) Post-translational regulation of cold acclimation response. Plant Sci 205:48–54

Burke MJ, Gusta LV, Quamme HA, Weiser CJ, Li PH (1976) Freezing and injury in plants. Annu Rev Plant Physiol 27(1):507–528

Buts K, Michielssens S, Hertog ML, Hayakawa E, Cordewener J, America AH, Nicolai BM, Carpentier SC (2014) Improving the identification rate of data independent label-free quantitative proteomics experiments on non-model crops: a case study on apple fruit. J Proteome 105:31–45

Bykova NV, Hoehn B, Rampitsch C, Hu J, Stebbing JA, Knox R (2011) Thiol redox-sensitive seed proteome in dormant and non-dormant hybrid genotypes of wheat. Phytochemistry 72(10):1162–1172

Cappadona S, Baker PR, Cutillas PR, Heck AJ, van Breukelen B (2012) Current challenges in software solutions for mass spectrometry-based quantitative proteomics. Amino Acids 43(3):1087–1108

Cheng L, Gao X, Li S, Shi M, Javeed H, Jing X, He G (2010) Proteomic analysis of soybean [*Glycine max* (L.) Meer.] seeds during imbibition at chilling temperature. Mol Breed 26(1):1–17

Cook D, Fowler S, Fiehn O, Thomashow MF (2004) A prominent role for the CBF cold response pathway in configuring the low-temperature metabolome of Arabidopsis. Proc Natl Acad Sci 101(42):15243–15248

Cui S, Huang F, Wang J, Ma X, Cheng Y, Liu J (2005) A proteomic analysis of cold stress responses in rice seedlings. Proteomics 5(12):3162–3172

Evans C, Noirel J, Ow SY, Salim M, Pereira-Medrano AG, Couto N et al (2012) An insight into iTRAQ: where do we stand now? Anal Bioanal Chem 404:1011–1027. https://doi.org/10.1007/s00216-012-5918-6

Fanucchi F, Alpi E, Olivieri S, Cannistraci CV, Bachi A, Alpi A, Alession M (2012) Acclimation increases freezing stress response of *Arabidopsis thaliana* at proteome level. Biochim Biophys Acta 1824(6):813–825

Flinn CL, Ashworth EN (1994) Blueberry flower-bud hardiness is not estimated by differential thermal analysis. J Am Soc Hortic Sci 119(2):295–298

Folgado R, Panis B, Sergeant K, Renaut J, Swennen R, Hausman JF (2013) Differential protein expression in response to abiotic stress in two potato species: *Solanum commersonii* Dun and *Solanum tuberosum* L. Int J Mol Sci 14(3):4912–4933

Franks F (1985) Complex aqueous systems at subzero temperatures. In: Properties of water in foods. Springer, Dordrecht, pp 497–509

Gao L, Yan X, Li X, Guo G, Hu Y, Ma W, Yan Y (2011) Proteome analysis of wheat leaf under salt stress by two-dimensional difference gel electrophoresis (2D-DIGE). Phytochemistry 72(10):1180–1191

Gharechahi J, Alizadeh H, Naghavi MR, Sharifi G (2014) A proteomic analysis to identify cold acclimation associated proteins in wild wheat *(Triticum urartu* L.). Mol Biol Rep 41(6):3897–3905

Gharechahi J, Sharifi G, Komatsu S, Salekdeh GH (2016) Proteomic analysis of crop plants under low temperature: a review of cold responsive proteins. In: Agricultural proteomics, vol 2. Springer, Cham, pp 97–127

Gillet LC, Navarro P, Tate S, Röst H, Selevsek N, Reiter L, Bonner R, Aebersold R (2012) Targeted data extraction of the MS/MS spectra generated by data-independent acquisition: a new concept for consistent and accurate proteome analysis. Mol Cell Proteomics 11(6):O111.016717

Gu X, Gao Z, Zhuang W, Qiao Y, Wang X, Mi L, Zhang Z, Lin Z (2013) Comparative proteomic analysis of rd29A:RdreB1BI transgenic and non-transgenic strawberries exposed to low temperature. J Plant Physiol 170(7):696–706

Hasanuzzaman M, Nahar K, Alam MM, Roychowdhury R, Fujita M (2013) Physiological, biochemical, and molecular mechanisms of heat stress tolerance in plants. Int J Mol Sci 14(5):9643–9684

Hashimoto M, Komatsu S (2007) Proteomic analysis of rice seedlings during cold stress. Proteomics 7(8):1293–1302

Heidarvand L, Amiri RM (2010) What happens in plant molecular responses to cold stress? Acta Physiol Plant 32(3):419–431

Higgins SJ, Hames BD (eds) (1999) Protein expression: a practical approach (No. 202). Oxford University Press, New York

Hlaváčková I, Vítámvás P, Šantrůček J, Kosová K, Zelenková S, Prášil IT, Kodíček M (2013) Proteins involved in distinct phases of cold hardening process in frost resistant winter barley (*Hordeum vulgare* L.) cv Luxor. Int J Mol Sci 14(4):8000–8024

Holman JD, Dasari S, Tabb DL (2013) Informatics of protein and posttranslational modification detection via shotgun proteomics. In: Proteomics for biomarker discovery. Humana Press, Totowa, pp 167–179

Kang YJ, Lee T, Lee J, Shim S, Jeong H, Satyawan D, Kim MY, Lee SH (2016) Translational genomics for plant breeding with the genome sequence explosion. Plant Biotechnol J 14(4):1057–1069

Körner C (2016) Plant adaptation to cold climates. F1000Research 5(F1000 Faculty Rev):2769

Kosová K, Vítámvás P, Planchon S, Renaut J, Vanková R, Prášil IT (2013) Proteome analysis of cold response in spring and winter wheat (*Triticum aestivum*) crowns reveals similarities in stress adaptation and differences in regulatory processes between the growth habits. J Proteome Res 12(11):4830–4845

Kwon SY, Jeong YJ, Lee HS, Kim JS, Cho KY, Allen RD, Kwak SS (2002) Enhanced tolerances of transgenic tobacco plants expressing both superoxide dismutase and ascorbate peroxidase in chloroplasts against methyl viologen-mediated oxidative stress. Plant Cell Environ 25(7):873–882

Lee DG, Ahsan N, Lee SH, Lee JJ, Bahk JD, Kang KY, Lee BH (2009) Chilling stress-induced proteomic changes in rice roots. J Plant Physiol 166(1):1–11

Levitt J (1980) Responses of plants to environmental stress, Volume 1. In: Chilling, freezing, and high temperature stresses. Academic Press, New York

Liu JY, Men JL, Chang MC, Feng CP, Yuan LG (2017) iTRAQ-based quantitative proteome revealed metabolic changes of Flammulina velutipes mycelia in response to cold stress. J Proteome 156:75–84

Matros A, Kaspar S, Witzel K, Mock HP (2011) Recent progress in liquid chromatography-based separation and label-free quantitative plant proteomics. Phytochemistry 72(10):963–974

Miura K, Furumoto T (2013) Cold signaling and cold response in plants. Int J Mol Sci 14(3):5312–5337

Nakaminami K, Matsui A, Nakagami H, Minami A, Nomura Y, Tanaka M, Morosawa T, Ishida J, Takahashi S, Uemura M, Shirasu K, Seki M (2014) Analysis of differential expression patterns of mRNA and protein during cold-and de-acclimation in Arabidopsis. Mol Cell Proteomics 13(12):3602–3611. https://doi.org/10.1074/mcp.M114.039081

Neilson KA, Mariani M, Haynes PA (2011) Quantitative proteomic analysis of cold-responsive proteins in rice. Proteomics 11(9):1697–1706

Owiti J, Grossmann J, Gehrig P, Dessimoz C, Laloi C, Hansen MB et al (2011) iTRAQ−based analysis of changes in the cassava root proteome reveals pathways associated with post−harvest physiological deterioration. Plant J 67:145–156. https://doi.org/10.1111/j.1365-313X.2011.04582.x

Pascovici D, Gardiner DM, Song X, Breen E, Solomon PS, Keighley T, Molloy MP (2013) Coverage and consistency: bioinformatics aspects of the analysis of multirun iTRAQ experiments with wheat leaves. J Proteome Res 12(11):4870–4881

Pearce AK, Humphrey TC (2001) Integrating stress-response and cell-cycle checkpoint pathways. Trends Cell Biol 11(10):426–433

Pearce RS, Fuller MP (2001) Freezing of barley studied by infrared video thermography. Plant Physiol 125(1):227–240

Picotti P, Clément-Ziza M, Lam H, Campbell DS, Schmidt A, Deutsch EW, Röst H, Sun Z, Rinner O, Reiter L, Shen Q (2013) A complete mass-spectrometric map of the yeast proteome applied to quantitative trait analysis. Nature 494(7436):266

Rinalducci S, Egidi MG, Karimzadeh G, Jazii FR, Zolla L (2012) Proteomic analysis of a spring wheat cultivar in response to prolonged cold stress. Electrophoresis 32(14):1807–1818

Rinalducci S, Egidi MG, Karimzadeh G, Jazii FR, Zolla L (2011) Proteomic analysis of a spring wheat cultivar in response to prolonged cold stress. Electrophoresis 32(14):1807–1818

Rosenzweig C, Parry ML (1994) Potential impact of climate change on world food supply. Nature 367(6459):133–138

Sánchez R, Flores A, Cejudo FJ (2006) Arabidopsis phosphoenolpyruvate carboxylase genes encode immunologically unrelated polypeptides and are differentially expressed in response to drought and salt stress. Planta 223(5):901–909

Sanghera GS, Wani SH, Hussain W, Singh NB (2011) Engineering cold stress tolerance in crop plants. Curr Genomics 12(1):30–43

Shi Y, Ding Y, Yang S (2018) Molecular regulation of CBF signaling in cold acclimation. Trends Plant Sci

Sobhanian H, Razavizadeh R, Nanjo Y, Ehsanpour AA, Jazii FR, Motamed N, Komatsu S (2010) Proteome analysis of soybean leaves, hypocotyls and roots under salt stress. Proteome Sci 8(1):19

Taylor NL, Heazlewood JL, Day DA, Millar AH (2005) Differential impact of environmental stresses on the pea mitochondrial proteome. Mol Cell Proteomics 4(8):1122–1133

Tian N, Wang J, Xu ZQ (2011) Overexpression of Na+/H+ antiporter gene AtNHX1 from Arabidopsis thaliana improves the salt tolerance of kiwifruit (Actinidia deliciosa). S Afr J Bot 77(1):160–169

Timperio AM, Egidi MG, Zolla L (2008) Proteomics applied on plant abiotic stresses: role of heat shock proteins (HSP). J Proteome 71(4):391–411

Uvackova L, Skultety L, Bekesova S, McClain S, Hajduch M (2013a) The MSE-proteomic analysis of gliadins and glutenins in wheat grain identifies and quantifies proteins associated with celiac disease and baker's asthma. J Proteome 93:65–73

Uvackova L, Skultety L, Bekesova S, McClain S, Hajduch M (2013b) MSE based multiplex protein analysis quantified important allergenic proteins and detected relevant peptides carrying known epitopes in wheat grain extracts. J Proteome Res 12(11):4862–4869

Virdi AS, Singh S, Singh P (2015) Abiotic stress responses in plants: roles of calmodulin-regulated proteins. Front Plant Sci 6:809

Walsh GM, Rogalski JC, Klockenbusch C, Kast J (2010) Mass spectrometry-based proteomics in biomedical research: emerging technologies and future strategies. Expert Rev Mol Med 12:e30

Wienkoop S, Baginsky S, Weckwerth W (2010) *Arabidopsis thaliana* as a model organism for plant proteome research. J Proteome 73(11):2239–2248

Xuan J, Song Y, Zhang H, Liu J, Guo Z, Hua Y (2013) Comparative proteomic analysis of the stolon cold stress response between the C4 perennial grass species *Zoysia japonica* and *Zoysia metrella*. PLoS One 8(9):e75705

Zhang M, Li G, Huang W, Bi T, Chen G, Tang Z, Sun W (2010) Proteomic study of *Carissa spinarum* in response to combined heat and drought stress. Proteomics 10(17):3117–3129

Zhang W, Zhang H, Ning L, Li B, Bao M (2016) Quantitative proteomic analysis provides novel insights into cold stress responses in Petunia seedlings. Front Plant Sci 7:136

Zheng B-B, Fang Y-N, Pan Z-Y, Sun L, Deng X-X, Grosser JW et al (2014) iTRAQ-based quantitative proteomics analysis revealed alterations of carbohydrate metabolism pathways and mitochondrial proteins in a male sterile *cybrid pummelo*. J Proteome Res 13:2998–3015. https://doi.org/10.1021/pr500126g

Chapter 8
What Can Small Molecules Tell Us About Cold Stress Tolerance in Plants?

Valentina Longo, Mohsen Janmohammadi, Lello Zolla, and Sara Rinalducci

8.1 Introduction

Suboptimal climatic conditions lead to abiotic stresses, which are the major factors causing significant decreases in agricultural productivity. It has been estimated that less than 5% of the global land area are unaffected by any environmental constraint (Van Velthuizen 2007). On the other hand, the world population is estimated to reach 10 billion by 2050, resulting in serious food shortages. In this context, it is important to develop stress tolerant crops to feed the increasing world population. Cold stress is one of the most important factors limiting plant growth, development and distribution, particularly in northern latitudes and at high altitudes (Yadav 2010). Freezing tolerance (FT) is a central factor influencing the yield and geographical distribution of crops. However, plants possess different abilities to overcome adverse environmental conditions. Some plant species can increase their degree of FT in response to chilling temperatures, a phenomenon known as "cold

---

Author contributed equally with all other contributors. Valentina Longo and Mohsen Janmohammadi

V. Longo
National Research Council, Institute for Microelectronics and Microsystems, Lecce, Italy
e-mail: valentina.longo@le.imm.cnr.it

M. Janmohammadi
Department of Plant Production and Genetics, Agriculture College, University of Maragheh, Maragheh, Iran
e-mail: jmohamad@alumni.ut.ac.ir

L. Zolla
Department of Science and Technology for Agriculture, Forestry, Nature and Energy (DAFNE), University of Tuscia, Viterbo, Italy
e-mail: zolla@unitus.it

S. Rinalducci (✉)
Department of Ecological and Biological Sciences (DEB), University of Tuscia, Viterbo, Italy
e-mail: sara.r@unitus.it

S. H. Wani, V. Herath (eds.), *Cold Tolerance in Plants*,
https://doi.org/10.1007/978-3-030-01415-5_8

acclimation" (CA; Thomashow 1999). Tolerance mechanisms differ between sudden freezing shock and freezing stress with prior CA. Predominantly, CA is induced by temperatures below 10 °C and by a short photoperiod in certain species of trees and grasses. To start this process, plants require days to weeks for full development. However, in some plants, FT can also be induced by other factors, such as osmotic stress and dehydration (Mäntylä et al. 1995; Li et al. 2002), salt stress (Ryu et al. 1995) and exogenous application of phytohormone abscisic acid (Li et al. 2003). The common denominator of all these events is a significant change in the metabolite content of plant cells.

Metabolism refers to life-sustaining biochemical phenomena that occur within plants. Plants endeavour to produce a variety of metabolites to attain biological balance and to ensure metabolic homeostasis. Metabolites are small molecules (from 50 to 1500 Da) contributing to metabolic reactions, which are essential for cellular function maintenance and growth (Aretz and Meierhofer 2016). The metabolome refers to the complete set of metabolites found within a biological sample and may include various chemical compounds such as amino acids, organic acids, nucleic acids, fatty acids, amines, sugars, vitamins, cofactors, pigments, antioxidants, etc. However, the metabolome is highly dynamic, time-dependent and sensitive to many environmental conditions (Kuehnbaum and Britz-McKibbin 2013; Aretz and Meierhofer 2016). Similarly, metabolites are extremely different (in polarity, charge, pKa, solubility, volatility, stability and reactivity), so that no single method can capture and analyse the entire metabolome (Johnson and Gonzalez 2012). The metabolome of an organism directly correlates with different pathways operating inside the cell, which, in turn, reflects the availability of corresponding genetic information. Since metabolites are the final products of a biological system in response to environmental changes or gene regulations, their evaluation can provide valuable information about the environmental conditions. Metabolomics is the term coined for essentially comprehensive, non-biased, high-throughput analyses of complex metabolite mixtures typical of plant extracts (Hall et al. 2002). The key benefit of metabolomics is the linking of metabolic networks to the underlying reaction pathway structure (Janmohammadi 2012; Hong et al. 2016). Therefore, metabolism evaluation has a number of advantages, such as clarifying connectivity maps of pathways, but we are still far from covering every part of the metabolome (Aretz and Meierhofer 2016).

The advantages may be partially due to the large number of metabolites within the cell. This metabolic affluence not only comes from the number of genes present, but is also a result of multiple substrate specificities for many enzymes (Hall et al. 2002), subcellular compartmentation and the occurrence of non-enzymic reactions. More than 50000 various compounds have been detected in plants (De Luca and St. Pierre 2000), and it is forecasted that the final figure for the plant kingdom will approach or even exceed 200000. Thus, metabolomics represents a considerable challenge for plant scientists.

Figure 8.1 shows the connections among "omics" sciences and emphasises how environmental effects impact the metabolome.

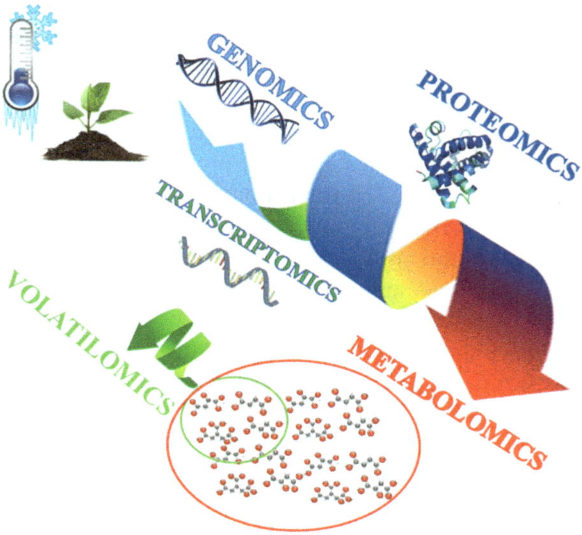

Fig. 8.1 Illustration of cold stress in plants and the possibility of assessing its effects on various levels through the "omics" techniques. Comparison shows the varying influence of the environment and physiology on the genome, proteome and the metabolome

Interestingly, a relatively new category in the "omics" arena is "volatilomics" (Achyuthan et al. 2017), which can be considered as a large branch of metabolomics, studying the volatile metabolic profile of an organism. The plant volatilome is composed of essential oils (EOs) and volatile organic compounds (VOCs), fed by different biosynthetic pathways and produced by plants, constitutively and/or after induction, as a defence strategy against biotic and abiotic stress (Maffei et al. 2011). Volatile organic compounds are defined as any organic compound with vapour pressures high enough under normal conditions to be vaporised (Dicke and Loreto 2010). They are released from leaves, flowers and fruits into the atmosphere and from roots into the soil, attracting pollinators, seed dispersers and other beneficial animals and microorganisms and serving as signals in plant-plant communication (Maffei 2010). Numerous plants emit VOCs in response to light and temperature changes or other abiotic stresses such as flooding or drought (Holzinger et al. 2000; Kreuzwieser et al. 2000). Thus, the detection of both high abundant metabolites and principal volatile compounds can help us to better understand how cold tolerance can be improved.

8.2 Acclimation Process

Depending on the species, studies have shown that acclimation proceeds in either one, two or three stages (Puhakainen 2004; Gusta et al. 2005; Herman et al. 2006; Guan 2014; Strimbeck et al. 2015). The first step includes initial acclimation to low temperature and concerns plant growth and upregulation of transcriptional factors. During this stage, plants attain tolerance to short frosts of −5 to −9 °C. Annual plants such as petunia, canola and oat show this type of acclimation (Puhakainen 2004). The second stage is in response to temperatures that approach 0 °C (Gusta et al. 2005; Herman et al. 2006). During this stage, cryoprotective compounds (such as proteins and sugars) are produced, and repair mechanisms develop. These substances are used to protect biological tissues from freezing damage (due to ice formation). Also, perennial and woody plants in boreal to arctic environments and on high mountains can survive prolonged exposure to temperatures below −40 °C and minimum temperatures below −60 °C, and laboratory tests have shown that many of these species can also survive immersion in liquid nitrogen at −196 °C (Gusta et al. 2005; Guan 2014; Strimbeck et al. 2015). These extremely hardy species survive sub-zero temperatures by tolerating freeze-induced desiccation. Such plants usually enter the third stage of acclimation, which may continue at mild freezing temperatures, in a process termed "sub-zero acclimation", resulting in a further increase in FT. This step is a combination of both biochemical and biophysical events (Gusta et al. 2005; Le et al. 2008). Winter-annual and perennial plants, such as winter cereals, grasses and trees, fall into this category, and some of them can tolerate temperatures as low as −196 °C for extended periods. However, it appears that ice formation inside the cells, triggered by an unusually rapid drop in temperature, is the cause of tissue death (Ninagawa et al. 2016), emphasising the importance of sub-zero acclimation and the changes made during this critical process. In woody plants, the CA process begins with a short-day (SD) length. Subsequent low and freezing temperatures are required to develop full frost tolerance. Leaves of silver birch (*Betula pendula*) are able to recognise and respond to both SD and low temperatures by increasing their frost tolerance (Puhakainen 2004). In general, it is proposed that the survival of these plants at extremely low temperatures can be due to the following mechanisms preventing irreversible injuries: (i) changes in lipid composition, which stabilise membranes at temperatures above the lipid phase transition temperature (−20 to −30 °C); (ii) production of high concentrations of oligosaccharides, promoting vitrification or high viscosity in the cytoplasm of freeze-dehydrated cells; and (iii) action of dehydrins, which bind membranes and further improve vitrification or act as lipid factor (stearically) to avoid deleterious membrane-membrane interactions (Guan 2014; Strimbeck et al. 2015). Although some general trends have been identified, there is not sufficient information about metabolome responses during the acclimation stage. Moreover, as a result of global warming, the vegetation suffers from repeated freeze-thaw cycles caused by more frequent short-term low temperatures induced by hail, snow or night frost. The metabolic changes in this process are also extremely important. Therefore, short-term freezing stress of plants should be investigated from the aspect of the metabolome.

During CA, osmotically active substances and osmoprotectants are accumulated, and the ratio of water to dry matter weight decreases (Prášil et al. 2001). Also, a change of intracellular calcium concentration is an early event during abiotic stresses (Lindberg et al. 2012). To achieve a comprehensive understanding for designing the breeding process, access to metabolomic information is essential. In addition, it is necessary to recognise the changes occurred during CA step by step at the metabolome level, from stimulus to perception and from perception to early signal transduction, which further eventually leads to dramatic alterations such as changes in plasma membrane fluidity, actin cytoskeleton rearrangement, cytosolic Ca^{2+} influx, protein phosphorylation, altered gene activity, increased formation of new protective gene products and secondary metabolites (Tuteja and Mahajan 2007; Solanke and Sharma 2008). Some compatible solutes, such as betaines, sugars (mannitol, sorbitol and trehalose), amino acids, flavonoids, polyols and polyamines, play crucial roles in the tolerance of plants against chilling stress. In addition, S-methylmethionine (SMM), an important intermediate compound in the sulphur metabolism, can be found in various quantities in the majority of plants (Liu et al. 2015; Rácz et al. 2008).

8.3 Metabolome Changes During Cold Acclimation

Cold stress can adversely affect crop productivity, quality and postharvest life, depending upon the degree of severity, growth stage and exposure duration. However, some plants are able to improve FT through the CA process. The freezing tolerance of such plants increases considerably after a period of exposure to low but nonfreezing temperatures (Thomashow 1999; Smallwood and Bowles 2002). Here, CA is accompanied by the activation of a large number of cold-responsive genes, which regulate metabolite concentrations (Koike et al. 2002). In contrast, spring crops or plants from warmer regions, such as rice and banana, are incapable of cold acclimation and often suffer significant damage and even death, upon exposure to chilling temperatures between 0 and 10 °C.

Some well-identified changes during CA are shown in Fig. 8.2.

Although recent advance in *Arabidopsis* genomics partially uncovered some important cold-responsive pathways during CA, such as the CBF (C-repeat binding factor) pathway (Cook et al. 2004; Lee and Thamashow 2012), we are far from a complete understanding of how the broad metabolite changes cause the increase of freezing and chilling tolerance. It has been recognised that CA is associated with an initial downregulation of photosynthesis, an extension of hormonal responses and the induction of the flavonoid metabolism. Evaluating the 1000 most significantly up- and downregulated transcripts in *Arabidopsis* during CA revealed a considerable over-representation of secondary metabolism, membrane transporters and primary metabolism among the upregulated transcripts, whereas photosynthesis, lipid metabolism, nucleotide metabolism, hormone and redox metabolism were among the downregulated transcripts (Hannah et al. 2005).

In the earliest steps of the acclimation, stress signals are recognised by sensors that are mainly located at the plasma membrane, resulting in the release or activation of various secondary messengers, such as calcium, reactive oxygen species (ROS) and inositol phosphates, which relay the stress signals and activate down-

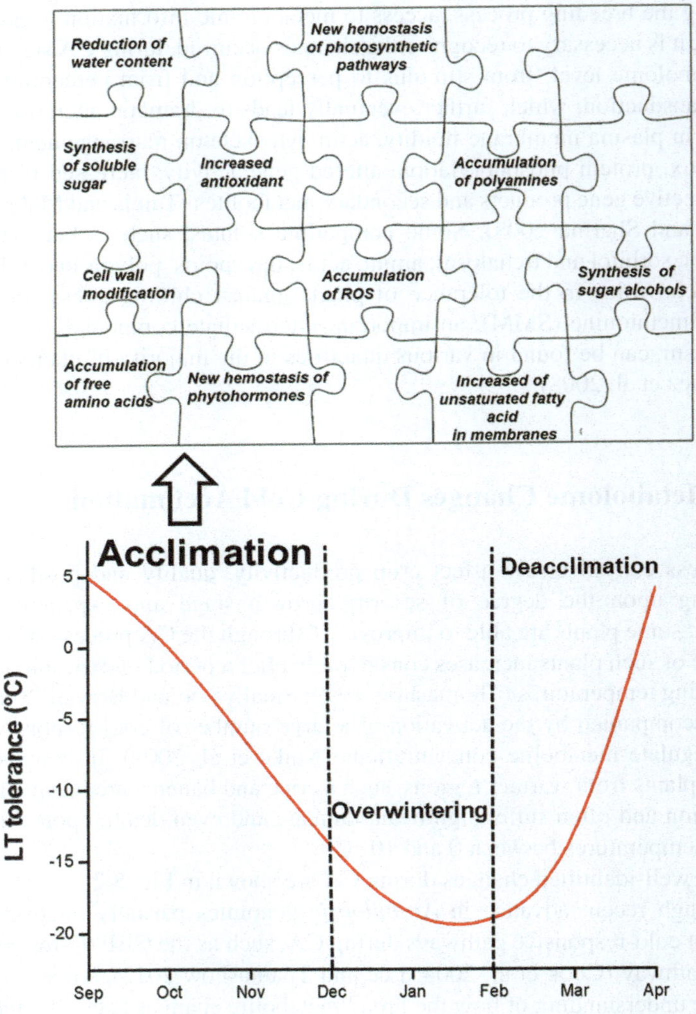

Fig. 8.2 Metabolic and physiological changes during cold acclimation in winter cereals

stream components, such as protein kinases and protein phosphatases (Ma et al. 2015). Cold responses are manifested by a range of morphological, physiological, biochemical and molecular changes. The complexity of the LT response makes it difficult to separate genes responsible for CA and cold hardiness from those associated with metabolic adjustments to cold condition. However, multiple mechanisms play a role in this context, including the synthesis of small metabolites with cryoprotective function, such as proline, sucrose and raffinose (Gilmour et al. 2000; Taji et al. 2002; Kaplan et al. 2007). These metabolites facilitate plant survival by stabilising membrane proteins and the phospholipid bilayer. Similarly, during CA, plants accumulate different low-molecular-weight compatible solutes to maintain cell osmotic pressure (Pirzadah et al. 2014), possibly also acting as antioxidants in cold stress responses. It is further believed that osmoregulation would be the best approach for abiotic stress tolerance (Bhatnagar-Mathur et al. 2008; Liu et al. 2013).

Carbohydrates play a key regulatory role in numerous vital processes of photosynthetic pathways, besides serving the energetic function, and are considered as influential signals which regulate plant metabolism and development. The alteration in the content or composition of water-soluble carbohydrates in response to abiotic stresses constitutes an important metabolic rearrangement in temperate grasses (Rao et al. 2011). Also, carbohydrates play an essential role in plant CA, but the communication between CA and sugar largely remains unknown. The metabolism of sugar is a complex and dynamic process controlled by numerous enzymes whose expression and activities change in response to LT stimulation (Yue et al. 2015). Studies have revealed that cold stress induces starch degradation and the main enzyme-coding genes involved in this pathway, including glucan water dikinase and beta-amylase, as well as its downstream genes, such as maltose transporter and disproportionating enzyme 2, are differentially regulated and expedite plant cold responses (Kaplan and Guy 2005; Li et al. 2011; Purdy et al. 2013). Furthermore, water-soluble polymeric sugars (such as fructans) are the major reserve carbohydrates, especially in temperate grasses. They are involved in the maintenance of the consistency and stabilisation of cell membranes to reduce water leakage (Hincha et al. 2000; Slewinski 2012), in the debarment of cell volume reduction by enhancing osmotic pressure and in freezing point depression (Krasensky and Jonak 2012). In addition, they play a role as antioxidants by scavenging ROS, thereby preventing cellular damage during abiotic stress conditions (Peshev et al. 2013). A possible scenario is that the fluctuation of fructan levels improves cold tolerance via a combination of different mechanisms. There is a close link between sugar and hormone signalling, regulating plant responses during CA (Pirzadah et al. 2014).

Polyamines (PAs), such as spermidine (a triamine), spermine (a tetramine) and their obligate precursor putrescine (a diamine), are ubiquitous polycationic aliphatic compounds which play a pivotal role in the regulation of plant development and the adaptation to environmental stress. They may also function as stress messengers in plant responses to different stress signals (Liu et al. 2007). It has been revealed that the PA concentration significantly increases during CA (Cuevas et al. 2008), and it appears that one of the ways by which PAs invigorate growth is the interaction with macromolecules (e.g. DNA, RNA, proteins) due to their polycationic structure (Yuki et al. 1996). According to previous studies, PAs play a role in the conformation maintenance and protection of DNA and in RNA stabilisation (D'Agostino et al. 2005; Liu et al. 2015). Furthermore, PAs affect cold tolerance by modulating ROS homeostasis due to their direct, or indirect, roles in regulating antioxidant systems or in suppressing ROS production (Liu et al. 2015).

8.4 Overwintering from the Perspective of the Metabolome

Freezing tolerance is dynamic and considerably affected by the developmental stage and variations in environmental temperature. In addition to CA, in the early months of autumn, winter-annual and perennial plants tend to increase FT through sub-zero acclimation during the winter season. However, the metabolic changes during this period have rarely been considered. Overwintering is the process by which plants pass through or wait out the winter season or pass through the period of the year in which harsh winter conditions (sub-zero temperatures, ice, snow, snowmold attacks) negatively affect the normal activity and make it difficult or near impossible to follow up the former trends. At such times, the growth of vegetative tissues and reproductive structures becomes minimal or ceases completely; however, plant responses may be considered a complex malady rather than a single reaction. For plants, overwintering often involves restricted water supplies and reduced light exposure. Overwintering in cold regions is associated with ice formation in extracellular and, in worse cases, intracellular spaces. Plant cells experience dehydration during freezing stress due to the presence of ice in extracellular spaces (Fujikawa 2016). Plasma membrane injury is mainly due to dehydration during the freeze-thaw cycle. Freezing-induced destabilisation of the plasma membrane involves different types of lesions (Thomashow 1999; Baumann 2017).

Changes in the accumulation of metabolites in buds are an important strategy for the overwintering of grapevine (Ershadi et al. 2016). Soluble carbohydrates accumulated in plants serve as cryoprotective compounds, which can prevent or slow ice crystal formation during winter (Karimi and Ershadi 2015). In addition to carbohydrates, trees significantly accumulate proline during the winter season, which induces osmotic adjustment, maintains turgor in dehydrating cells and allows them to tolerate dehydrative stresses caused by ice formation (Zhang et al. 2012). The decrease in bud water content is highly related to the development of FT and is thought to contribute to an increased ability to supercool (Valle 2002; Ershadi et al. 2016). Therefore, it is feasible to use alterations in these compatible materials in trees exposed to FT stress as indicators to assess FT in different grapevine cultivars. Ershadi et al. (2016) have reported a substantial decrease in soluble carbohydrates in spring, which may be due to their allocation to processes such as cell growth or conversion to starch during the deacclimation stage (Morin et al. 2007).

The genetic systems responsible for overwintering of cereals are developmentally regulated and induced by specific environmental conditions (Fowler and Limin 2004). In winter plants, vernalisation requirement is an important adaptive feature that delays the initiation of floral primordia by postponing the transition from the vegetative to the reproductive phase (Mahfoozi et al. 2006). For these reasons, the full expression of FT is only revealed at the point of vernalisation fulfilment, which is genetically controlled and affected by combinations of time and environmental cues (Fowler and Limin 2004). Hence, vernalisation requirement can affect the amount and duration of metabolite accumulation in the cell. It has also been noted that each plant has a distinct temperature threshold to initiate the acclimation process and plants differ in maintaining the maximum FT during winter (Fowler 2008; Fowler et al. 2014). In winter cereals, the maximum of FT level can be maintained when crown temperatures remain near or below freezing. At this stage, the acquired LT tolerance is rapidly lost once acclimated plants are exposed to temperatures above the acclimation threshold. Temperature fluctuations during winter can significantly reduce and/or erase FT; however, the process of LT acclimation can be restarted by exposing plants that are still in the vegetative stage to temperature points below the threshold (Mahfoozi et al. 2001; Janmohammadi 2010; Janmohammadi et al. 2012a; Janmohammadi et al. 2015).

It appears that cold-induced metabolites reach maximum levels during overwintering (Pasandi and Janmohammadi 2015). In this context, the proline content, photosynthetic pigments and hydrogen peroxide (H_2O_2) considerably increased during overwintering in spring and winter wheat. However, the increasing rate and the time for which they remained at the maximum levels were significantly different between genotypes (Hosseini et al. 2016). These results demonstrate that both genetic and environmental factors, by developmental regulation of metabolite accumulation, play important roles in creating FT. On the other hand, reports indicate that the highest concentrations of cold-induced metabolites, such as proline and antioxidants, were recorded at the point of vernalisation fulfilment (Janmohammadi et al. 2012b; Hosseini et al. 2016). In general, studies have indicated that cellular levels of osmolytes, such as proline, glycine betaine (GB) and sugars such as trehalose, fructans and sugar alcohols, which play a pivotal role in osmotic adjustment during water-deficit situation, were induced by CA and remained at high levels during overwintering (Bhandari and Nayyar 2014). The close relationship between the vegetative/reproductive transition and the start of a decline in FT demonstrates the regulatory influence of developmental genes on LT-tolerant genes and metabolic changes in winter cereals. The observed dependence on the stage of phenological development and the significant correlations of cold-induced metabolites with FT indicate that they are co-regulated responses.

Regarding the defensive roles of small metabolites under cold stress conditions (osmotic adjustments, cryoprotection, ROS scavenging), studies have shown that their accumulation during CA increases, and, particularly in winter cereals, the highest level of accumulation is achieved at the maximum FT point (double ridge stage). However, this does not mean that the production of these metabolites decreases as plants enter the reproductive stage. Indeed, many of them are essential for active plant growth, so that the production of small metabolites, especially amino acids, PAs, pigments and sugars, strongly increases with the transition into the reproductive phase. However, because of their high consumption in growth and development processes, their accumulation in the cell significantly decreases.

During overwintering, accumulated metabolites may also serve as cryoprotectants to save the cellular metabolism by protecting the integrity of membranes and cellular organelles, maintaining the redox potential and components of vital pathways, saving the photosynthetic machinery and also acting as partial antioxidants (Bhandari and Nayyar 2014).

8.5 Volatile Organic Compounds (VOCs) and Abiotic Stress

Abiotic stresses are rarely experienced singularly; rather, they often occur in combination. Furthermore, only few plant responses are stress-specific (Vickers et al. 2009). Alterations in cell pathways are translated in synthesis or emanation of particular volatile metabolites or in expression-level changes of VOC patterns. In addition to ubiquitous stress-elicited volatiles, emissions of a number of specific volatiles are often induced from stressed plants, including mono- and sesquiterpenes and homoterpenes, with the blend of volatiles depending on the stress type (Niinemets 2010). Some of these induced compounds are signal molecules, while others are specific products of stress response pathways. Often, an important consequence of stress is an excess production of ROS, which can induce cell damage through oxidation of nucleic acids, proteins and lipids, leading to programmed cell death. Plants have a complex network of antioxidant compounds; among them, non-volatile compounds are the most known ones. However, in reality, volatile antioxidants also play a role in defence strategies. In this category, the principal actors are volatile isoprenoids (Vickers et al. 2009).

Isoprenoids are a large and various group of organic compounds. Their skeletons are composed of five carbons, differentially connected and with different modifications. Their production depends on two separate but linked pathways: (i) the cytosolic mevalonic acid (MVA) pathway and (ii) the plastidic 2-C-methyl-D-erythritol 4-phosphate (MEP) pathway. Volatile isoprenoids are lipophilic and low-molecular-weight compounds (under 300 Da) (Fig. 8.3). Only hemiterpenes (isoprene and methylbutenol, monoterpenes, sesquiterpenes and some diterpenes) have sufficient vapour pressure to volatilise at biological temperatures (Dudareva et al. 2006). Plants invest enormously in terms of carbons and energy to produce volatile isoprenoids. In fact, being volatile, their loss is irretrievable and represents about 1–2% of the photosynthetic carbon fixation (Sharkey and Yen 2001). Under stress conditions, although photosynthesis is severely inhibited, isoprenoid emission is sustained (Brilli et al. 2007), suggesting that this VOC confers benefits to the plant,

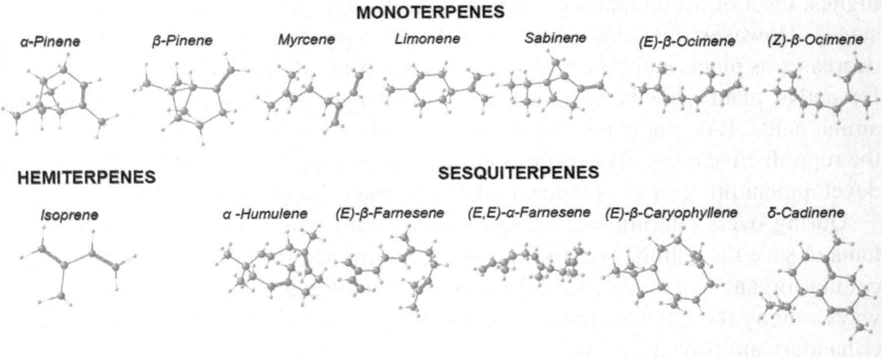

Fig. 8.3 Chemical structures of volatile isoprenoids with antioxidant properties

although for many volatiles, these benefits remain obscure. Currently, the most investigated VOC is isoprene. There is strong evidence that volatile isoprenoids protect photosynthesis under thermal and oxidative stress conditions (Kuzuyama et al. 1998).

8.5.1 VOC Release During Cold Stress

Cold stress, such as all biotic and abiotic stresses, induces changes in the volatile compound released by plants. For example, emissions of short-chained alcohols and aldehydes (C6), so-called green leaf volatiles, products of the lipoxygenase (LOX) pathway, constitute one of the first stress responses of plants (Loreto et al. 2006). For cold stress, significantly enhanced LOX emissions have not been observed until a severe stress induced by temperatures of −7 °C and less (Copolovici et al. 2012). Monoterpene emissions are similar in heat and cold stress. In fact, in both cases, membrane permeability is altered, resulting in plastid pH conditions more suitable for terpene synthesis. In particular, for cold stress treatments, (E)-β-ocimene, a monoterpene characteristically induced after stress, was quantitatively correlated with stress temperature (Copolovici et al. 2012). In the literature, there is more information about heat than cold. Since VOCs partition between the gas and liquid phases within plants, according to their Henry's law constant, and since the equilibrium between the two phases is determined by temperature (Niinemets et al. 2004; Harley 2013), high temperature shifts the balance towards the gaseous phase and thereby allows a greater release of VOCs. The detection of high levels of volatile compounds in cold-stressed plants, such as (E)-β-ocimene in *Solanum lycopersicum* (Copolovici et al. 2012), reflects a drastic situation that exceeds the limit imposed by Henry's law.

8.6 Main Actors in the Cold-Responsive Small-Molecule World

Basically, from the metabolomic point of view, a large number of molecules are formed during cold stress. However, three types of compounds are implicated in this process: (i) compounds involved in the acclimation (e.g. antioxidants or osmoprotectants, cryoprotectants, metabolism modulators, cytoskeleton reorganisers), (ii) by-products derived from cold stress due to alterations in growth conditions (e.g. oxidised compounds, γ-aminobutyric acid, PPi, organic acids, glutamate, malondialdehyde) and (iii) signal transduction molecules acting as mediators of the CA response (e.g. Ca^{2+}, inositol phosphates, ROS, ABA). A large range of physiological and biochemical alterations occur during CA and are shown in Table 8.1. The most notable changes include growth reduction or cessation, decreased tissue water content, transient

Table 8.1 Some of the identified metabolite changes during cold acclimation

Metabolites	Trends	Plant	Organs	References
Proline	↑	*Brachypodium distachyon*	Leaf	Colton-Gagnon et al. (2014)
Tyrosine (↑) aspartic acid (↑) glutamine (↓) arginine (↑) ornithine (↑) citrulline (↑)	↑↓	*Bluegrass, Arabidopsis*	Crown	Dionne et al. (2001) Cook et al. (2004)
Sucrose, raffinose	↑↓	*Arabidopsis, sweet cherry, tea plant*	Shoot	Rekarte-Cowie et al. (2008) Turhan and Ergin (2012) Yue et al. (2015)
Trehalose	↑	*Rice*	Shoot	Pramanik and Imai (2005)
Fructooligosaccharides	↑	*Ryegrass*	Shoot	Abeynayake et al, (2015)
γ-Aminobutyric acid	↑	*Barley and wheat*	Leaf	Mazzucotelli et al. (2006)
Mannitol, glycerol, sorbitol	↓	*Delphinium*	Root	Ogasawara et al. (2001);
Putrescine, spermidine	↑	*Arabidopsis*	Seedlings	Kasukabe et al. (2004); Cuevas et al. (2008); Alcázar et al. (2011)
Glutamate	↑	*Spring wheat*	Shoot	Kovács et al. (2012)
Longer-chain triacylglycerols (↑) digalactosyldiacylglycerol (↓) Monogalactosyldiacylglycerol (↑) glucosylceramide (↓) ceramide (↑)	↑↓	*Arabidopsis*	Shoot	Degenkolbe et al. (2012)
Flavonoids	↑	*Arabidopsis*	Leaf	Catalá et al. (2011) Schulz et al. (2016)
Organic acid (malate, fumarate, succinate, citrate)	↑	*Arabidopsis*	Leaf	Nagler et al. (2015)
Hydrogen peroxide superoxide anion singlet oxygen	↑	*Barley and Arabidopsis*	Shoot	Dai et al. (2009) Chinnusamy et al. (2006) Baek and Skinner (2012)
Water soluble antioxidants (ascorbate, glutathione)	↑	*Arabidopsis*	Shoot	Dai et al. (2009)
Membrane-associated antioxidants (α-tocopherol, β-carotene and ubiquinone)	↑	*Arabidopsis*	Shoot	Maeda et al. (2006)

increase in abscisic acid, changes in the membrane lipid composition and accumulation of some small metabolites (Xin and Browse 2000). In the following sections, the role of some key cold-induced metabolites is described in detail.

8.6.1 The Role of Betaine During Cold Stress

Chemically, a betaine is any neutral chemical compound with a positively charged cationic functional group, such as a quaternary ammonium or a phosphonium cation. Biologically, betaine is involved in methylation reactions and homocysteine detoxification. Glycine betaine (*N*,*N*,*N*-trimethylglycine; GB) was the first betaine discovered and is an amino acid derivative in plants. It has a zwitterionic nature, and its natural accumulation is associated with abiotic stress tolerance in various organisms (Kishitani et al. 1994; Giri 2011). Glycine betaine is categorised as one of the important organic compatible solutes, which are defined as small molecules that are highly soluble in water and uniformly neutral with respect to the perturbation of cellular functions, even when present at high concentrations (Wani et al. 2013). The levels of accumulated GB are generally correlated with the extent of the stress tolerance and vary considerably among plant species and organs (Chen and Murata 2008). Many plants accumulate low-molecular-weight compounds with cryoprotective activity in response to low temperatures, and GB is one of these compounds (Giri 2011; Wani et al. 2016). Although most abiotic stresses result in increased GB levels, the stimulating effect of cold stress is higher than that of other strains. In this context, Xing and Rajashekar (2001) have reported that ABA-induced GB levels in *Arabidopsis thaliana* were considerably smaller than those resulting from CA treatment, indicating that the accumulation of GB under cold stress occurs independently of ABA (Kurepin et al. 2015). However, in winter-annual plants, GB accumulation during CA is associated with improved freezing tolerance (Janmohammadi et al. 2012b; Hosseini et al. 2016). Among these, rice is the only cereal that does not accumulate GB naturally (Shirasawa et al. 2006). Apart from plants, the osmoprotective functions of GB have also been shown for lower organisms, e.g. in *Vibrio anguillarum* (Ma et al. 2017) and *Listeria monocytogenes* (Ells and Hansen 2011). In higher plants, GB is biosynthesised from choline by a two-step oxidation reaction (Fig. 8.4). An early precursor of choline, ethanolamine, is converted to monomethyl ethanolamine by the addition of a methyl group from S-adenosylmethionine (SAM). The monomethyl ethanolamine conversion to choline occurs in the cytosol, and choline is then transported by a choline transporter to the chloroplasts, where the final biosynthesis phase of GB occurs in a two-step synthesis, involving choline monooxygenase (CMO) and betaine aldehyde dehydrogenase (BADH), whose expression and activity are significantly induced by abiotic stress (Kurepin et al. 2015; Wani et al. 2016). This metabolic route has been extensively studied, and the creation of transgenic plants through manipulation of this pathway has resulted in improved tolerance to environmental stresses, including cold stress (Yadav 2010). We know that low-temperature stress considerably

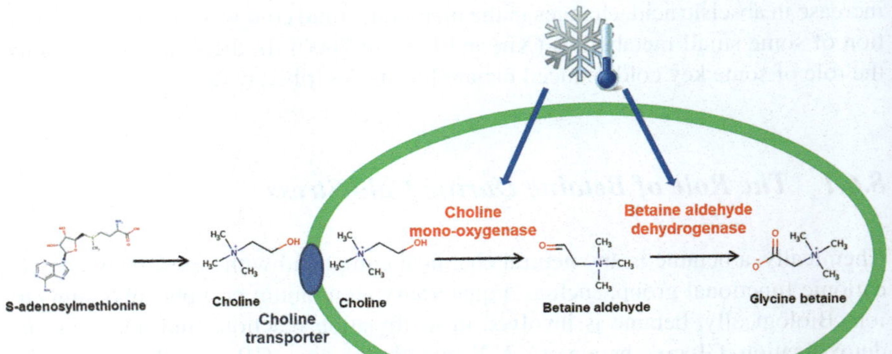

Fig. 8.4 Proline biosynthesis in plants and the effect of cold stress on enzymes involved in this pathway

induces the production of ROS, which causes damage to cellular components or prevents numerous activities by disturbing the cellular redox balance (Awasthi et al. 2015). A moderate increase in the levels of H_2O_2 in GB-transgenic plants (choline monooxygenase) possibly activates H_2O_2-inducible protective mechanisms (Park et al. 2004), suggesting an interrelationship between ROS signalling and GB accumulation (Megha et al. 2014). On the other hand, GB also can be synthesised from glycine. This pathway involves three successive *N*-methylations of glycine (Heldt and Piechulla 2010) and is catalysed by two methyltransferases, glycine sarcosine methyltransferase (GSMT) and sarcosine dimethylglycine methyltransferase (SDMT) or dimethylglycine methyltransferase (DMT), with partially overlapping substrate specificities (Lai et al. 2006). However, the presence of the genes encoding the just mentioned enzymes does not mean that GB can accumulate in the plant (Giri 2011), which is clearly visible in rice plants.

Although GB can accumulate in various organelles and intracellular spaces, its most protective performance occurs in chloroplasts. In these organelles, GB plays an imperative role in the amendment and protection of the thylakoid membrane, thereby maintaining photosynthetic efficiency and ROS scavenging (dos Reis et al. 2012). In this regard, increased GB levels are also associated with an increase in photosynthesis rates, which is considered to be one of the major mechanisms for survival under stress conditions (Kurepin et al. 2015; Wani et al. 2016). In transgenic rice with chloroplast-targeted GB accumulation, the protection of the photosynthetic machinery against salt and cold stress was better than in plants with cytosolic GB accumulation, even though GB accumulation was fivefold higher in the latter plants (Sakamoto and Murata 2002). However, at lower concentrations, GB effectively contributes to stabilising the quaternary structures of enzymes and complex proteins (Chen and Murata 2008). Besides, GB can protect plants from adverse effects of a range of abiotic stresses by maintaining the water balance between plant cells and the environment, acting as an osmotic regulator.

8.6.2 Sugar Changes Under Cold Stress

Sugars are small soluble carbohydrates, including monosaccharides (mostly fructose and glucose), disaccharides (sucrose, trehalose) and oligosaccharides (raffinose and stachyose), which play an essential role during CA (Morsy et al. 2007; Janmohammadi et al. 2012b; Yue et al. 2015). During cold exposure, numerous carbohydrates accumulate, and a new metabolic homeostasis evolves (Nägele et al. 2012). Starch hydrolysis via the β-amolytic pathway represents the predominant pathway of starch degradation in leaves under normal growth conditions, but it may also be involved in stress situation (Krasensky and Jonak 2012). The amount of carbohydrate polymers, such as starch, significantly decreases under cold stress, and, conversely, the concentrations of soluble sugars increase during the CA of overwintering plants (Liu et al. 2018). Sugars not only act as primary energy and carbon sources but also as signalling molecules in different physiological processes. They function as cryoprotectants, osmolytes and stabilisers of proteins and enzymes and are involved in maintaining membrane integrity and ROS mitigation (Bhandari and Nayyar 2014; Tarkowski and Van den Ende 2015), thus playing multifunctional roles in plant growth and development (Ruan 2014). In contrast to the common opinion, cold-induced accumulation of sugars does not occur in the vacuole but rather in the cytoplasm, thereby protecting various membrane systems (Partelli et al. 2010).

It seems that there is a close relationship between cold tolerance and sugar metabolism. For example, cold-tolerant chickpea genotypes have higher activities of enzymes involved in carbohydrate metabolism (Kumar et al. 2011). There is a significant positive correlation between sugar content and FT, and in one study, induced FT wheat seeding was lost after short-term deacclimation at control temperatures, and this change was associated with a large reduction in sugar content (Janmohammadi 2010).

Trehalose, also known as mycose or tremalose, is a natural alpha-linked reduced disaccharide formed by an α,α-1,1-glucoside bond between two α-glucose units (Elbein et al. 2003; Eastmond et al. 2002). The regulatory roles of trehalose-6-phosphate, a precursor of trehalose, in sugar metabolism, growth and development in plants have already been revealed (Eastmond et al. 2002). Trehalose has an excellent capacity to protect membranes and proteins from degradation (Magazù et al. 2012). Recent studies have shown that the genes encoding trehalose-6-phosphate synthases (TPSs) and trehalose-6-phosphate phosphatases (TPPs) are responsive to cold stress in *Arabidopsis* (Iordachescu and Imai 2008; Sah et al. 2016). Stachyose, a tetrasaccharide with two galactose, one glucose and one fructose molecules, increased gradually in coffee genotypes under low temperatures (Partelli et al. 2010). In particular, the authors showed that sucrose concentration slightly increased during cold stress, which was concomitantly associated with an increase in glucose and fructose levels (Partelli et al. 2010). Interestingly, this trend has also been reported by Tapernoux-Lüthi et al. (2004) in other plants. Probably, this status was not due to an increased production of precursors but rather to their lower use by the cellular metabolism, since photosynthesis was significantly reduced and was negligible under cold stress. Also, the reduction of sucrose in some plants might be due to the inhibition of sucrose-P synthase (SPS). In fact, the inability of fully activating SPS is among the factors that contribute to a poor photosynthetic performance and a higher cold sensitivity. Cold stress may result in SPS inhibition and reduced sucrose synthesis. In this regard, in a previous study, the exposure of either spring or winter wheat cultivars to low-temperature conditions was associated with feedback-limited photosynthesis, as indicated by a 50–60% reduction in CO_2 assimilation rates, a twofold lower ATP/ADP ratio and a threefold lower electron transport rate than in the control plants grown at 20 °C (Savitch et al. 1997). In contrast to low-temperature stress, gradual CA of cereals results in an increase in photosynthetic capacity (Bravo et al. 2007). Increased CO_2 assimilation rates following CA are associated with increases in the activities of RuBisCO, stromal and cytosolic fructose-1,6-bisphosphatase (Hurry et al. 1995; Bravo et al. 2007) as well as SPS (Du and Nose 2002).

As previously mentioned, the *Arabidopsis* (CBF C-repeat binding factor) cold response pathway plays a fundamental role during cold hardening, the process whereby plants increase their freezing tolerance by experiencing low, nonfreezing temperatures (Lee and Thomashow 2012). Galactinol and raffinose, sugars that are synthesised through the action of the CBF-targeted gene *AtGolS3*, as well as melibiose, fructose, glucose, inositol and sucrose, sugars associated with the raffinose pathway, all occurred at lower levels in freezing-susceptible *Arabidopsis* plants (Cvi-1) as compared with freezing-tolerant (Ws-2) plants (Cook et al. 2004). However, transgenic plants constitutively overexpressing *CBFs* showed higher induction of the *STZ/ZAT10* zinc finger transcription factor gene, which appears to repress genes involved in photosynthesis and carbohydrate metabolism and thus to reduce the growth of these transgenic plants (Maruyama et al. 2004). Still, with regard to cytosolic biosynthesis of the cold-responsive sugars, any changes in subcellular concentrations of these cryoprotectants will depend on the number and activity of the transporters (Tarkowski and Van den Ende 2015). Previous experiments have revealed that the activity and/or expression of sugar transporters may be regulated by signalling, affecting the subcellular distribution and the overall cellular homeostasis of sugars, and this may be tightly linked to the cellular redox homeostasis (Schneider and Keller 2009; Klemens et al. 2013; Guo et al. 2014). Both sucrose and hexoses play dual functions in gene regulation, as exemplified by the upregulation of growth-related genes and the downregulation of stress-related genes (Rosa et al. 2009). Sucrose and glucose either act as substrates for cellular respiration or as osmolytes to maintain cell homeostasis, while fructose does not seem to be related to osmoprotection but rather associated to secondary metabolite synthesis (Gupta et al. 2013).

8.6.3 Polyols

Chemically, polyols are alcohols containing multiple hydroxyl groups; they are reduction products of aldoses or ketoses. Compared to simple sugars and disaccharides, polyols have been underestimated under cold stress. Polyols are involved in the following processes: (i) the enhancement of photosynthesis, supporting redox control, osmotic adjustment, storage and translocation of carbon; (ii) the stabilisation of macromolecules, and (iii) the scavenging of hydroxyl radicals, thereby preventing oxidative damage of membranes and enzymes (Shen et al. 1999; Karakas 2001; Krasensky and Jonak 2012). Polyols have less feedback inhibitory effects on photosynthesis when compared with other sugars because of the additional cytosolic sink for photosynthetically fixed CO_2, used for polyol synthesis. Most likely, this makes them suitable candidates for genetic manipulation and breeding to improve plant tolerance under abiotic stress, such as low temperatures. Plants transgenic for polyol-related genes have increased stress tolerance (dos Reis et al. 2012). Sorbitol, myo-inositol, D-ononitol, mannitol-1-phosphate and galactinol are the main polyols that play protective functions under stress conditions. The biosynthesis of the mentioned polyols is catalysed by sorbitol-6-phosphate dehydrogenase, inositol-1-phosphate synthase, inositol methyltransferase, mannitol-1-phosphate dehydrogenase and galactinol synthase, respectively; an overexpression of these enzymes can significantly improve plant tolerance under cold stress (for more details, see Krasensky and Jonak 2012).

There is a close relationship between mannitol, sorbitol and their corresponding metabolic enzymes, supporting the idea that sorbitol synthesis should respond positively to stress conditions through increased mRNA levels (Karakas 2001). Furthermore, a common approach to accumulate sorbitol is to submit the plant to stress conditions that induce sorbitol transporter genes (dos Reis et al. 2012).

8.6.4 The Role of Polyamines in Cold Tolerance

Polyamines (PAs) are small aliphatic, positively charged and organic molecules with two or more primary amino groups – NH_2; they are ubiquitous in plants. There is growing evidence that PAs are not only involved in numerous physiological processes of plant growth and development, but also play important roles in modulating the defence responses of plants to diverse environmental stresses (Martin-Tanguy 2001; Kader et al. 2011; Liu et al. 2015). Polyamines are well known for their anti-senescence due to their acid-neutralising and antioxidant properties as well as their membrane- and cell wall-stabilising abilities (Zhao and Yang 2008; Gill and Tuteja 2010). They also play fundamental roles in protein synthesis, cell cycle regulation, ion channel regulation, signal transduction and gene expression, but most likely, there are numerous other, albeit not yet revealed, metabolic processes in which PAs are protagonists (Takahashi and Kakehi 2010).

It appears that cellular PAs typically accumulate in plants under both short- and long-term abiotic stress conditions, and this is consistent with the possibility of their dual effects, i.e. being protectors from, as well as perpetrators of, stress damage to cells (Minocha et al. 2014; Kader et al. 2011). In addition, PA accumulation at high levels can be valuable in the nitrogen balance, and its distribution into multiple pathways can reduce ammonia toxicity (Minocha et al. 2014; Liu et al. 2015).

There are three major PAs in plants, i.e. putrescine, spermidine and spermine, although other types, such as cadaverine, can also be present. It seems that putrescine accumulation under cold stress is essential for proper cold hardening and survival at freezing temperatures, because *Arabidopsis* mutants defective in putrescine biosynthesis (*adc1*, *adc2*) display reduced freezing tolerance compared to wild-type plants (Cuevas et al. 2008). It has also been suggested that putrescine acts as a signalling molecule, interacting with ABA-dependent signalling pathways involved in cold stress.

Polyamines play a crucial role in controlling ROS homeostasis via two ways. Firstly, they may inhibit the autoxidation of metals, which in turn impairs the supply of electrons for the generation of ROS (Shi et al. 2010). They may also directly act as antioxidants and scavenge ROS, although at present, there is no evidence for this mechanism. Secondly, PAs may induce antioxidant systems. Several studies have shown that exogenous application of spermidine and putrescine increased the activities of superoxide dismutase, guaiacol peroxidase and catalase when compared to non-treated chilled plants (Abavisani et al. 2013; Zhang et al. 2009; Kader et al. 2011). These results strongly suggest that PAs act as stimulants of the antioxidant machinery to prevent chilling injury through counteracting oxidative stress. Besides, high levels of PAs may improve FT by acting as direct ROS scavengers or binding to antioxidant enzyme molecules to scavenge free radicals (Minocha et al. 2014; Liu et al. 2015). Accordingly, exogenous application of spermidine effectively suppresses electrolyte leakage in wheat (Kader et al. 2011), cucumber (Zhang et al. 2009) and dragonhead (Abavisani et al. 2013).

Cold stress can also amplify the glycolysis pathway, which is associated with the production of methylglyoxal (MG), a toxic compound (Nahar et al. 2015). The recognised cytotoxic effects of methylglyoxal include clogged membrane structure and function, increased ion leakage, deterioration of ultrastructural molecules and organelles, DNA damage or even cell death (Yadav 2010). Plants are equipped with the glyoxalase pathway to reduce the toxicity of methylglyoxal. Exogenous application of spermidine on mung bean seedling under cold stress noticeably reduced methylglyoxal toxicity by improving glyoxalase cycle components and by maintaining osmoregulation, water status and a significantly increased seedling growth (Nahar et al. 2015). These researchers also revealed that PA application could accelerate the activity of the ascorbate-glutathione (AsA-GSH) cycle as an important antioxidant defence system in plants.

However, the response of different PA compounds to cold stress differs between various plants species and organs, suggesting that PA accumulation is influenced by different factors, such as plant species, stress tolerance capacity, stress type and condition as well as the physiological status of the examined tissues/organs. It also indicates the existence of complex PA dynamics under abiotic stress, which may explain the contradictory results reported from various studies. The size of the PA pool can be correlated with the stress tolerance capacity, further underlining the significance of PAs in providing protection against stresses (Liu et al. 2015). There is a positive correlation between the number of amino groups in PAs and their efficiency under stress conditions, and spermidine and spermine, in contrast to putrescine, play more important roles in counteracting abiotic stress since they contain one and two additional primary amino groups ($-NH_2$), respectively.

8.6.5 Amino Acid Changes at Low Temperatures

Plants under suboptimal conditions tend to accumulate proline and other amino acids. Accumulated amino acids in plants play various roles ranging from osmosis regulation, modulation of ion transport, disturbance of stomatal opening (Rai and Sharma 1991) and detoxification of heavy metals. Amino acids also affect the synthesis and activities of some enzymes, gene expression and redox homeostasis (Rai 2002). Based on the literature, proline, glutamic acid, glutamine, aspartic acid, asparagine, threonine, serine, leucine and histidine are the most responsive amino acids to abiotic stress (Kovács et al. 2012; Janmohammadi 2014). As an example, evaluation of amino acid concentrations in three annual bluegrass ecotypes during cold and sub-zero acclimation showed that acclimation could significantly increase the concentrations of glutamine, asparagine, arginine, tyrosine, proline and total amino acids (Dionne et al. 2001). Specifically, plants from northern climates accumulated higher levels of amino acids than those from other ecotypes.

Previous studies have suggested a positive correlation between proline accumulation and plant stress (Hayat et al. 2012). In *Arabidopsis*, CA considerably increases proline concentrations (Kaplan et al. 2007), and proline directly safeguards key cellular macromolecules (especially membrane lipids and proteins such as enzymes (Verbruggen and Hermans 2008)) and contributes to the protection of cellular functions by scavenging ROS. Furthermore, it induces the expression of stress-responsive genes which possess proline-responsive elements (Ashraf and Foolad 2007). Free proline facilitates the preservation of an osmotic equilibrium between the symplast and the apoplast, thus aiding in preventing low-temperature damage by maintaining the functional integrity of the plasma membrane (Hosseini et al. 2016). Studies have also shown that proline can alter the activities of antioxidant enzymes (de Campos et al. 2011).

Proline is predominantly synthesised from glutamate via D1-pyrroline-5-carboxylate (P5C) by two successive reductions that are catalysed by P5C synthetase (P5CS), the rate-limiting enzyme for the biosynthetic pathway in higher plants, and P5C reductase (P5CR). Cold acclimation induces the gene expression or activity of enzymes involved in proline biosynthesis (Megha et al. 2014). In this context, the proline biosynthetic gene *P5CS2* (pyrroline-5-carboxylate synthetase 2), as a

CBF-targeted gene, is significantly induced by CA in *Arabidopsis* plants (Liang et al. 2013). In addition, the downregulation of proline dehydrogenase increases proline levels (dos Reis et al. 2012).

The role of some amino acid derivatives is also significant under stress conditions. For example, gamma-aminobutyric acid (GABA), which is synthesised from glutamine by a single decarboxylation reaction, plays a versatile role in plants and, together with proline, is one of the most common osmolytes accumulated by plant cells in response to stress (Mazzucotelli et al. 2006; Okumoto et al. 2016). Accumulation of GABA improves the quality of the crop yield and contributes to defence reactions but also occurs under non-stress condition in specific tissues. Gamma-aminobutyric acid is further catabolised within the mitochondria through a two-step pathway called the "GABA shunt", which bypasses a part of the TCA cycle. This pathway plays a crucial role in carbon metabolism, especially at specific growth stages or under adverse conditions (Okumoto et al. 2016). In addition, GABA can also function as a signalling molecule by directly modulating some membrane channels. Furthermore, GABA transport seems to act in regulating the C/N balance. However, the increase in free amino acids can be a precondition for increasing the biosynthesis of cold-responsive proteins. In this regard, COR15, WCS120 and late embryogenesis abundant-like proteins (LEA) are noteworthy.

8.6.6 Metabolome Changes in the Chloroplast as a Key Organ

Chloroplasts are not only the place of photosynthesis (photoreduction of carbon, nitrogen and sulphur) but are central hubs in plant metabolism (Neuhaus and Emes 2000). In maize inbreds adapted to low temperatures, a reduced sensitivity of photosynthesis to cold has been observed (Singh 2014). Due to the localisation of many key cycles and metabolic processes in chloroplasts, it appears that cold stress significantly changes the metabolome of these organelles, making chloroplasts the major sites of cold stress damages. Simple sugars, arginine, ornithine, glutamine and GB are the most important metabolites produced in the chloroplasts. Carbohydrate metabolism has a greater instantaneous low-temperature sensitivity than other components of photosynthesis (Theocharis et al. 2012; Janmohammadi and Mahfoozi 2013). Cold stress decreases carbon and other enzymatic assimilation processes, creating a greater imbalance because light absorption is largely temperature-insensitive (Huner et al. 1998). In one study, cold stress resulted in long-term changes in the organisation of the photosynthetic apparatus in winter plants, with a decrease in the number of functional photosystem (PS) II reaction centres, a loss of light-harvesting chlorophylls (Chls) and a disrupted formation of the large thylakoid protein complex involved in LHC II, PS II and PS I (Ensminger et al. 2006). Also, in the seedling stage, the impact of cold stress sharply increases the concentration of the accessory pigments (Chl-b and carotenoids) when compared to Chl-a, probably to increase photon capture (Adam and Murthy 2014). Since chlorophyll concentrations can be used as good indicators for susceptibility to chilling injury, Janmohammadi (2010) claimed that there was a positive correlation between chlorophyll content and LT tolerance in winter wheat and ray. However, CA decreases the susceptibility to photoinhibition by causing several metabolic alterations and producing changes at the chloroplast level, which may restore the energy balance. Cold-induced tolerance includes increases in enzyme activities (e.g. from the Calvin cycle and sugar metabolism), reinforcement of energy dissipation mechanisms and antioxidative molecules (Adam and Murthy 2014).

Xanthophylls are the typical yellow pigments of leaves. Although they are not considered photosynthetic pigments per se, the xanthophylls (notably violaxanthin, antheraxanthin and zeaxanthin) help to protect the photosystems, and their abundance increases at low temperatures (Ivanov et al. 2006). Xanthophylls have structural roles and act as natural antioxidants, quenching triplet Chl and singlet oxygen, which are potentially harmful to the chloroplast (Han et al. 2010).

Finally, low temperatures also have a significant impact on the synthesis of flavonoids in plants. Flavonoids are a major class of secondary metabolites, constituting about 5–10% of the known secondary products in plants ranging from bryophytes to angiosperms, and are synthesised from phenylalanine via the polypropanoid pathway. They have been classified into six subclasses: (i) flavones (luteonin, apigenin, tangeritin), (ii) flavonols (quercetin, kaempferol, myricetin, isorhamnetin, pachypodol), (iii) flavanones (hesteretin, naringenin, eriodictyol), (iv) flavan-3-ols (catechins and epicatechins), (v) isoflavones (genistein, daidzein, glycitein), (vi) anthocyanidin compounds (cyanidin, delphinidin, malvidin, pelargonidin, peonidin, petunidin) and other common flavonoid groups, including aurones, xanthones and condensed tannins (Samanta et al. 2011).

Flavonoids are synthesised through the phenylpropanoid pathway, which is controlled by key enzymes including phenylalanine ammonia-lyase (PAL) and chalcone synthase (Sharma et al. 2007). Depending upon the range of low temperatures to which plants are subjected, CA leads to remarkable increases in PAL activity, resulting in the accumulation of flavonoids in leaves and stems (Stefanowska et al. 2002). Flavonoids may play functional roles in plant CA and FT and, especially quercetin, may scavenge ROS and act as potent antioxidants under cold stress. At sub-zero temperatures, large amounts of water are removed from the cell into intercellular ice crystals; in these circumstances, flavonoids are expected to partition even more strongly into the lipid phase of cell membranes, thus stabilising them (Korn et al. 2008). However, the regulation of cold-induced photosynthetic processes is complex and requires further studies.

8.6.7 The Role of Small Molecules in Cold Stress Signalling

A common signal transduction pathway initiates with signal perception, which is associated with alteration of plasma membrane fluidity. This step is followed by the generation of second messengers (e.g. inositol phosphates, nitric oxide and ROS), which can modulate intracellular Ca^{2+} levels, often starting a protein phosphorylation cascade that finally targets proteins directly involved in cellular protection or transcription factors controlling specific sets of stress-regulated genes (Hakeem et al. 2014). This signal is then transduced downstream and results in DNA modifications (Pirzadah et al. 2014). For instance, altered levels of DNA methylation appear to encourage proline accumulation via the upregulation of P5CS and ornithine-d-aminotransferase (OAT) enzymes. Thus proline, as a stress-responsive amino acid, acts as a signal of stress memory stably carried to the next generation (Kishor et al. 2014). Due to the role of proline in redox adjustment, its accumulation may also affect ROS signalling. In addition, some second messengers, such as nitric oxide, can significantly affect the biosynthesis of small metabolites. Nitric oxide mediates the post-translational alterations of specific proteins (through cysteine S-nitrosylation or tyrosine nitration), eventually increasing the scavenging capacity by the activation of superoxide dismutase and decreasing photosynthesis by RuBisCO inhibition. Besides direct effects on protein function, nitric oxide signalling regulates COR gene expression by induction or repression of the genes involved in proline metabolism (Kazemi-Shahandashti and Maali-Amiri 2018). To conclude, various signalling pathways, comprising ABA, jasmonic acid, auxin, ROS, MAPK, Ca^{2+} receptors, second messengers and so on, act together to modulate plant responses to cold stress, resulting in metabolite accumulation. In such a situation, combining the feedback effects on signalling pathways is not far from the mind.

8.7 Conclusions and Outlook for Future Research

Plant physiologists and plant molecular biologists have always been interested in mechanisms involved in plant tolerance to cold and in the ways in which plants may react to withstand damage following stress. Metabolome studies reveal that there are different categories of cold-responsive metabolites. Cold-responsive metabolites have cryoprotective activity and may also act as scavengers of free radicals, cellular redox modulators, osmolytes, stabilisers of proteins and enzymes, ameliorātives of membrane integrity, regulators of gene expression and the cell cycle as well as correctors of ion channel functions. They are also involved in signal transduction, carbon translocation and maintenance of cell homeostasis. However, it is important to also consider volatile metabolites, which, although they have not been extensively studied in this field, play a prominent, albeit often underestimated, role in these processes, which is particularly the case for isoprenoids. Volatile organic compounds are small lipophilic molecules and the first to cross membranes, therefore representing one of the primordial signals of plants both of stress perception and defence mechanisms.

Most metabolic adjustments often occur on extremely short timescales. The accumulation of osmoprotectants (including GB, sugars, PAs, amino acids, nitrogen-containing compounds and by-products of key cycles), changes in the lipid membrane profile as well as photosynthetic acclimation along with extensive reprogramming at the molecular level, help temperate plants acquire FT. The increase of free proline content in plants subjected to cold stress has been reported extensively, and there are proven relationships between proline accumulation during CA and FT. Trehalose is an important osmoprotectant found in minute amounts in plants, although it plays a major role in plant cell metabolism associated with abiotic stress tolerance, which has been investigated with the use of genetic engineering tools in crop plants. On the other hand, dehardening may have considerable effects on the plant metabolome. Plant growth and dehardening rate are both temperature-dependent. Evaluation of different dehardening trends can provide valuable information for plant breeding processes. Moreover, temperature has a decisive effect on the chloroplast and impairs the biosynthesis of several metabolites due to the down-regulation of gene expression and the abundance of several enzymes involved in their metabolism. On the other hand, gradual CA can increase the accumulation of some cryoprotectant and beneficial metabolites in these organelles. However, combining quantitative metabolite profiling with subcellular compartmentation analysis will be more efficient for improving the identification of mechanisms involved in cold tolerance. Conventional breeding methods had limited success in improving the FT of important crop plants, involving interspecific or intergeneric hybridisation. Metabolome profiling during cold hardening, along with full genome profiling and transgenic plant analyses, can provide a deep insight of the complex mechanisms that operate under freezing stress. Furthermore, for abundant metabolites, nontargeted metabolomics has the potential to capture information about relative levels of substrates and products for reversible rate-limiting enzymatic reactions.

References

Abavisani A, Khorshidi M, Sherafatmandjour A (2013) Interaction between cold stress and polyamine on antioxidant properties in dragonhead. Int J Agric Crop Sci 5(21):2555

Abeynayake SW, Etzerodt TP, Jonavičienė K, Byrne S, AspT BB (2015) Fructan metabolism and changes in fructan composition during cold acclimation in perennial ryegrass. Front Plant Sci 6:329

Achyuthan KE, Harper JC, Manginell RP, Moorman MW (2017) Volatile metabolites emission by in vivo microalgae-an overlooked opportunity? Meta 31(3):7

Adam S, Murthy SDS (2014) Effect of cold stress on photosynthesis of plants and possible protection mechanisms. In: Approaches to plant stress and their management. Springer, New Delhi, pp 219–226

Alcázar R, Cuevas JC, Planas J, Zarza X, Bortolotti C, Carrasco P, Altabella T (2011) Integration of polyamines in the cold acclimation response. Plant Sci 180(1):31–38

Aretz I, Meierhofer D (2016) Advantages and pitfalls of mass spectrometry based metabolome profiling in systems biology. Int J Mol Sci 17(5):632

Ashraf M, Foolad M (2007) Roles of glycine betaine and proline in improving plant abiotic stress resistance. Environ Exp Bot 59(2):206–216

Awasthi R, Bhandari K, Nayyar H (2015) Temperature stress and redox homeostasis in agricultural crops. Front Environ Sci 3:11

Baek KH, Skinner DZ (2012) Production of reactive oxygen species by freezing stress and the protective roles of antioxidant enzymes in plants. J Agric Chem Environ 1(1):34–40

Baumann K (2017) Stress responses: membrane-to-nucleus signals modulate plant cold tolerance. Nat Rev Mol Cell Biol 18(5):276

Bhandari K, Nayyar H (2014) Low temperature stress in plants: an overview of roles of cryoprotectants in defense. In: Physiological mechanisms and adaptation strategies in plants under changing environment. Springer, New York, pp 193–265

Bhatnagar-Mathur P, Vadez V, Kiran K, Sharma KK (2008) Transgenic approaches for abiotic stress tolerance in plants: retrospect and prospects. Plant Cell Rep 27:411–424

Bravo LA, Saavedra-Mella FA, Vera F, Guerra A, Cavieres LA, Ivanov AG, Huner NPA, Corcuera LJ (2007) Effect of cold acclimation on the photosynthetic performance of two ecotypes of Colobanthus quitensis (Kunth) Bartl. J Exp Bot 58(13):3581–3590

Brilli F, Barta C, Fortunati A, Lerdau M, Loreto F, Centritto M (2007) Response of isoprene emission and carbon metabolism to drought in white poplar (*Populus alba*) saplings. New Phytol 175(2):244–254

Catalá R, Medina J, Salinas J (2011) Integration of low temperature and light signaling during cold acclimation response in *Arabidopsis*. Proc Natl Acad Sci 108(39):16475–16480

Chen TH, Murata N (2008) Glycinebetaine: an effective protectant against abiotic stress in plants. Trends Plant Sci 13(9):499–505

Chinnusamy V, Zhu J, Zhu JK (2006) Gene regulation during cold acclimation in plants. Physiol Plant 126(1):52–61

Colton-Gagnon K, Ali-Benali MA, Mayer BF, DionneR BA, Do Carmo S, Charron JB (2014) Comparative analysis of the cold acclimation and freezing tolerance capacities of seven diploid *Brachypodium distachyon* accessions. Ann Bot 113(4):681–693

Cook D, Fowler S, Fiehn O, Thomashow MF (2004) A prominent role for the CBF cold response pathway in configuring the low-temperature metabolome of *Arabidopsis*. Proc Natl Acad Sci U S A 101(42):15243–15248

Copolovici L, Kännaste A, Pazouki L, Niinemets U (2012) Emissions of green leaf volatiles and terpenoids from *Solanum lycopersicum* are quantitatively related to the severity of cold and heat shock treatments. J Plant Physiol 169(7):664–672

Cuevas JC, López-Cobollo R, Alcázar R, Zarza X, Koncz C, Altabella T, Salinas J, Tiburcio AF, Ferrando A (2008) Putrescine is involved in Arabidopsis freezing tolerance and cold acclimation by regulating abscisic acid levels in response to low temperature. Plant Physiol 148(2):1094–1105

D'Agostino L, di Pietro M, Di Luccia A (2005) Nuclear aggregates of polyamines are supramolecular structures that play a crucial role in genomic DNA protection and conformation. FEBS J 272(15):3777–3787

Dai F, Huang Y, Zhou M, Zhang G (2009) The influence of cold acclimation on antioxidative enzymes and antioxidants in sensitive and tolerant barley cultivars. Biol Plant 53(2):257–262

de Campos MKF, de Carvalho K, de Souza FS, Marur CJ, Pereira LFP, Bespalhok Filho JC, Vieira LGE (2011) Drought tolerance and antioxidant enzymatic activity in transgenic 'Swingle'citrumelo plants over-accumulating proline. Environ Exp Bot 72(2):242–250

De Luca V, St Pierre B (2000) The cell and developmental biology of alkaloid biosynthesis. Trends Plant Sci 5(4):168–173

Degenkolbe T, Giavalisco P, Zuther E, Seiwert B, Hincha DK, Willmitzer L (2012) Differential remodeling of the lipidome during cold acclimation in natural accessions of *Arabidopsis thaliana*. Plant J 72(6):972–982

Dicke M, Loreto F (2010) Induced plant volatiles: from genes to climate change. Trends Plant Sci 15(3):115–117

Dionne J, Castonguay Y, Nadeau P, Desjardins Y (2001) Amino acid and protein changes during cold acclimation of green-type annual bluegrass (L,) ecotypes. Crop Sci 41(6):1862–1870

dos Reis SP, Lima AM, de Souza CRB (2012) Recent molecular advances on downstream plant responses to abiotic stress. Int J Mol Sci 13(7):8628–8647

Du YC, Nose A (2002) Effects of chilling temperature on the activity of enzymes of sucrose synthesis and the accumulation of saccharides in leaves of three sugarcane cultivars differing in cold sensitivity. Photosynthetica 40(3):389–395

Dudareva N, Negre F, Nagegowda DA, Orlova I (2006) Plant volatiles: recent advances and future perspectives. Crit Rev Plant Sci 25:417–440

Eastmond PJ, Van Dijken AJ, Spielman M, Kerr A, Tissier AF, Dickinson HG, Jones JD, Smeekens SC, Graham IA (2002) Trehalose-6-phosphate synthase 1 which catalyses the first step in trehalose synthesis is essential for *Arabidopsis* embryo maturation. Plant J 29(2):225–235

Elbein AD, Pan YT, Pastuszak I, Carroll D (2003) New insights on trehalose: a multifunctional molecule. Glycobiology 13(4):17R–27R

Ells TC, Hansen LT (2011) Increased thermal and osmotic stress resistance in *Listeria monocytogenes* 568 grown in the presence of trehalose due to inactivation of the phosphotrehalase-encoding gene treA. Appl Environ Microbiol 77(19):6841–6851

Ensminger I, Busch F, Huner N (2006) Photostasis and cold acclimation: sensing low temperature through photosynthesis. Physiol Plant 126(1):28–44

Ershadi A, Karimi R, Mahdei KN (2016) Freezing tolerance and its relationship with soluble carbohydrates proline and water content in 12 grapevine cultivars. Acta Physiol Plant 38(1):2

Fowler DB, Limin AE (2004) Interactions among factors regulating phenological development and acclimation rate determine low-temperature tolerance in wheat. Ann Bot 94(5):717–724

Fowler DB, Byrns BM, Greer KJ (2014) Overwinter low-temperature responses of cereals: analyses and simulation. Crop Sci 54(6):2395–2405

Fowler DB (2008) Cold acclimation threshold induction temperatures in cereals. Crop Sci 48:1147–1154

Fujikawa S (2016) Plant responses to freezing. In eLS, Wiley, Ltd (Ed.), Chichester

Gill SS, Tuteja N (2010) Polyamines and abiotic stress tolerance in plants. Plant Signal Behav 5(1):26–33

Gilmour SJ, Sebolt AM, Salazar MP, Everard JD, Thomashow MF (2000) Overexpression of the Arabidopsis CBF3 transcriptional activator mimics multiple biochemical changes associated with cold acclimation. Plant Physiol 124(4):1854–1865

Giri J (2011) Glycinebetaine and abiotic stress tolerance in plants. Plant Signal Behav 6(11):1746–1751

Guan X (2014) Physiology of cold acclimation and deacclimation responses of cool-season grasses: Carbon and hormone metabolism, Doctoral dissertation, University of Massachusetts Amherst. 174

Guo WJ, Nagy R, Chen HY, Pfrunder S, Yu YC, Santelia D, Martinoia E (2014) SWEET17 a facilitative transporter mediates fructose transport across the tonoplast of *Arabidopsis* roots and leaves. Plant Physiol 164(2):777–789

Gupta K, Dey A, Gupta B (2013) Plant polyamines in abiotic stress responses. Acta Physiologiae Plantarum 35(7):2015–2036

Gusta LV, Trischuk R, Weiser CJ (2005) Plant cold acclimation: the role of abscisic acid. J Plant Growth Regul 24(4):308–318

Hakeem KR, Rehman RU, Tahir I (2014) Plant signaling: understanding the molecular crosstalk. Springer, New Delhi

Hall R, Beale M, Fiehn O, Hardy N, Sumner L, Bino R (2002) Plant metabolomics the missing link in functional genomics strategies. Plant Cell 14(7):1437–1440

Han H, Gao S, Li B, Dong XC, Feng HL, Meng QW (2010) Overexpression of violaxanthin de-epoxidase gene alleviates photoinhibition of PSII and PSI in tomato during high light and chilling stress. J Plant Physiol 167(3):176–183

Hannah MA, Heyer AG, Hincha DK (2005) A global survey of gene regulation during cold acclimation in *Arabidopsis thaliana*. PLoS Genet 1(2):e26

Harley PC (2013) The roles of Stomatal conductance and compound volatility in controlling the emission of volatile organic compounds from leaves. In: Niinemets Ü, Monson R (eds) Biology controls and models of tree volatile organic compound emissions, tree physiology, vol 5. Springer, Dordrecht

Hayat S, Hayat Q, Alyemeni MN, Wani AS, Pichtel J, Ahmad A (2012) Role of proline under changing environments: a review. Plant Signal Behav 7(11):1456–1466

Heldt HW, Piechulla B (2010) Plant biochemistry. Academic Press, London

Herman EM, Rotter K, Premakumar R, Elwinger G, Bae R, Ehler-King L, Chen S, Livingston DP III (2006) Additional freeze hardiness in wheat acquired by exposure to −3 °C is associated with extensive physiological, morphological, and molecular changes. J Exp Bot 57(14):3601–3618

Hincha DK, Hellwege EM, Heyer AG, Crowe JH (2000) Plant fructans stabilize phosphatidylcholine liposomes during freeze-drying. Eur J Biochem 267(2):535–540

Holzinger R, Sandoval-Soto L, Rottenberger S, Crutzen PJ, Kesselmeier J (2000) Emissions of volatile organic compounds from *Quercus ilex L,* measured by proton-transfer-reaction mass spectrometry under different environmental conditions. J Geophys Res 105:20573–20579

Hong J, Yang L, Zhang D, Shi J (2016) Plant metabolomics: an indispensable system biology tool for plant science. Int J Mol Sci 17(6):767

Hosseini M, Maali-Amiri R, Mahfoozi S, Fowler DB, Mohammadi R (2016) Developmental regulation of metabolites and low temperature tolerance in lines of crosses between spring and winter wheat. Acta Physiol Plant 38(4):87–98

Huner NP, Öquist G, Sarhan F (1998) Energy balance and acclimation to light and cold. Trends Plant Sci 3(6):224–230

Hurry VM, Strand A, Tobiaeson M, Gardestrom P, Oquist G (1995) Cold hardening of spring and winter wheat and rape results in differential effects on growth carbon metabolism and carbohydrate content. Plant Physiol 109(2):697–706

Iordachescu M, Imai R (2008) Trehalose biosynthesis in response to abiotic stresses. J Integr Plant Biol 50(10):1223–1229

Ivanov AG, Krol M, Sveshnikov D, Malmberg G, Gardeström P, Hurry V, Oquist G, Huner NP (2006) Characterization of the photosynthetic apparatus in cortical bark chlorenchyma of *Scots pine*. Planta 223(6):1165–1177

Janmohammadi M (2010) Study of interrelationship between vegetative/reproductive transition stage and cold induced proteins expression using proteomics analysis in wheat grown under field conditions. P.hD. University of Tehran Faculty of Agricultural Sciences and Engineering. Tehran 236 pp.

Janmohammadi M (2012) Metabolomic analysis of low temperature responses in plants. Curr Opin Agric 1(1):1–6

Janmohammadi M, Enayati V, Sabaghnia N (2012a) Impact of cold acclimation, de-acclimation and re-acclimation on carbohydrate content and antioxidant enzyme activities in spring and winter wheat. Icel Agric Sci 25:3–11

Janmohammadi M, Mahfoozi S, Tvakkol-Afshari R (2012b) Influence of vegetative/reproductive transition on proline and protein accumulation and expression of freezing tolerance in wheat cultivars grown in temperate and cold climates. Agric For 58(3):67–84

Janmohammadi M, Mahfoozi S (2013) Regulatory network of gene expression during the development of frost tolerance in plants. Curr Opin Agric 2(1):11–19

Janmohammadi M (2014) Identification of leaf proteins differentially accumulated during the overwintering in frost-tolerant winter wheat. Curr Opin Agric 3(1):26–29

Janmohammadi M, Zolla L, Rinalducci S (2015) Low temperature tolerance in plants: changes at the protein level. Phytochemistry 117:76–89

Johnson CH, Gonzalez FJ (2012) Challenges and opportunities of metabolomics. J Cell Physiol 227:2975–2981

Kader DZA, Saleh AA, Elmeleigy SA, Dosoky NS (2011) Chilling-induced oxidative stress and polyamines regulatory role in two wheat varieties. J Taibah Univ Sci 5:14–24

Kaplan F, Guy CL (2005) RNA interference of *Arabidopsis* betaamylase8 prevents maltose accumulation upon cold shock and increases sensitivity of PSII photochemical efficiency to freezing stress. Plant J 44:730–743

Kaplan F, Kopka J, Sung DY, Zhao W, Popp M, Porat R, Guy CL (2007) Transcript and metabolite profiling during cold acclimation of *Arabidopsis* reveals an intricate relationship of cold-regulated gene expression with modifications in metabolite content. Plant J 50(6):967–981

Karakas B (2001) The role of sorbitol synthesis in photosynthesis of peach (*Prunus persica*). Doctoral dissertation uga. Athens Georgia

Karimi R, Ershadi A (2015) Role of exogenous abscisic acid in adapting of 'Sultana' grapevine to low-temperature stress. Acta Physiol Plant 37(8):1–11

Kasukabe Y, He L, Nada K, Misawa S, Ihara I, Tachibana S (2004) Overexpression of spermidine synthase enhances tolerance to multiple environmental stresses and up-regulates the expression of various stress-regulated genes in transgenic Arabidopsis thaliana. Plant Cell Physiol 45(6):712–722

Kazemi-Shahandashti SS, Maali-Amiri R (2018) Global insights of protein responses to cold stress in plants: signaling, defence, and degradation. J Plant Physiol 226:123–135

Kishitani S, Watanabe K, Yasuda S, Arakawa K, Takabe T (1994) Accumulation of glycinebetaine during cold acclimation and freezing tolerance in leaves of winter and spring barley plants. Plant Cell Environ 17(1):89–95

Kishor K, Polavarapu B, Sreenivasulu N (2014) Is proline accumulation per se correlated with stress tolerance or is proline homeostasis a more critical issue? Plant Cell Environ 37(2):300–311

Klemens PA, Patzke K, Deitmer J, Spinner L, Le Hir R, Bellini C, Chardon F, Krapp A, Neuhaus HE (2013) Overexpression of the vacuolar sugar carrier AtSWEET16 modifies germination growth and stress tolerance in *Arabidopsis*. Plant Physiol 163(3):1338–1352

Koike M, Okamoto T, Tsuda S, Imai R (2002) A novel plant defensin-like gene of winter wheat is specifically induced during cold acclimation. Biochem Biophys Res Commun 298(1):46–53

Korn M, Peterek S, Mock HP, Heyer AG, Hincha DK (2008) Heterosis in the freezing tolerance and sugar and flavonoid contents of crosses between *Arabidopsis thaliana* accessions of widely varying freezing tolerance. Plant Cell Environ 31(6):813–827

Kovács Z, Simon-Sarkadi L, Vashegyi I, Kocsy G (2012) Different accumulation of free amino acids during short-and long-term osmotic stress in wheat. Sci World J 2012:216521

Krasensky J, Jonak C (2012) Drought salt and temperature stress-induced metabolic rearrangements and regulatory networks. J Exp Bot 63(4):1593–1608

Kreuzwieser J, Ku¨hnemann F, Martis A, Urban W, Rennenberg H (2000) Diurnal pattern of acetaldehyde emission by flooded poplar trees. Physiol Plant 108:79–86

Kuehnbaum NL, Britz-McKibbin P (2013) New advances in separation science for metabolomics: resolving chemical diversity in a post-genomic era. Chem Rev 113:2437–2468

Kumar S, Malik J, Thakur P, Kaistha S, Sharma KD, Upadhyaya HD, Berger JD, Nayyar H (2011) Growth and metabolic responses of contrasting chickpea (*Cicer arietinum L.*) genotypes to chilling stress at reproductive phase. Acta Physiol Plant 33(3):779–787

Kurepin LV, Ivanov AG, Zaman M, Pharis RP, Allakhverdiev SI, Hurry V, Hüner NP (2015) Stress-related hormones and glycinebetaine interplay in protection of photosynthesis under abiotic stress conditions. Photosynth Res 126(2–3):221–235

Kuzuyama T, Shimizu T, Takahashi S, Seto H (1998) Fosmidomycin a specific inhibitor of 1-deoxy-D-xylulose 5-phosphate reductoisomerase in the nonmevalonate pathway for terpenoid biosynthesis. Tetrahedron Lett 39:7913–7916

Lai MC, Wang CC, Chuang MJ, Wu YC, Lee YC (2006) Effects of substrate and potassium on the betaine-synthesizing enzyme glycine sarcosine dimethylglycine N-methyltransferase from a halophilic methanoarchaeon Methanohalophilus portucalensis. Res Microbiol 157(10):948–955

Le MQ, Engelsberger WR, Hincha DK (2008) Natural genetic variation in acclimation capacity at sub-zero temperatures after cold acclimation at 4 C in different *Arabidopsis thaliana* accessions. Cryobiology 57(2):104–112

Lee CM, Thomashow MF (2012) Photoperiodic regulation of the C-repeat binding factor (CBF) cold acclimation pathway and freezing tolerance in Arabidopsis thaliana. Proc Natl Acad Sci 109(37):15054–15059

Li C, Junttila O, Heino P, Palva ET (2003) Different responses of northern and southern ecotypes of Betula pendula to exogenous ABA application. Tree Physiol 23:481–487

Li C, Puhakainen T, Welling A, Viherä-Aarnio A, Ernstsen A, Junttila O, Heino P, Palva ET (2002) Cold acclimation in silver birch (*Betula pendula*). Development of freezing tolerance in different tissues and climatic ecotypes. Physiol Plant 116:478–488

Li T, Xu SL, Oses-Prieto JA, Putil S, Xu P, Wang RJ, Li KH, Maltby DA, An LH, Burlingame AL, Deng ZP, Wang ZY (2011) Proteomics analysis reveals post-translational mechanisms for cold induced metabolic changes in *Arabidopsis*. Mol Plant 4:361–374

Liang X, Zhang L, Natarajan SK, Becker DF (2013) Proline mechanisms of stress survival. Antioxid Redox Signal 19(9):998–1011
Lindberg S, Kader MA, Yemelyanov V (2012) Calcium signalling in plant cells under environmental stress. In: Environmental adaptations and stress tolerance of plants in the era of climate change. Springer, New York, pp 325–360
Liu JH, Kitashiba H, Wang J, Ban Y, Moriguchi T (2007) Polyamines and their ability to provide environmental stress tolerance to plants. Plant Biotechnol 24:117–126
Liu W, Yu K, He T, Li F, Zhang D, Liu J (2013) The low temperature induced physiological responses of *Avena nuda L.*, a cold-tolerant plant species. Sci World J 2013:658793
Liu JH, Wang W, Wu H, Gong X, Moriguchi T (2015) Polyamines function in stress tolerance: from synthesis to regulation. Front Plant Sci 6:13
Liu W, Su J, Li S, Lang X, Huang X (2018) Non-structural carbohydrates regulated by season and species in the subtropical monsoon broad-leaved evergreen forest of Yunnan Province, China. Sci Rep 8(1):1083
Loreto F, Barta C, Brilli F, Nogues I (2006) On the induction of volatile organic compound emissions by plants as consequence of wounding or fluctuations of light and temperature. Plant Cell Environ 29:1820–1828
Ma Y, Dai X, Xu Y, Luo W, Zheng X, Zeng D, Pan Y, Lin X, Liu H, Zhang D, Xiao J, Guo X, Xu S, Niu Y, Jin J, Zhang H, Xu X, Li L, Wang W, Qian Q, Ge S, Chong K (2015) COLD1 confers chilling tolerance in rice. Cell 160(6):1209–1221
Ma Y, Wang Q, Gao X, Zhang Y (2017) Biosynthesis and uptake of glycine betaine as cold-stress response to low temperature in fish pathogen *Vibrio anguillarum*. J Microbiol 55(1):44–55
Maeda H, Song W, Sage TL, DellaPenna D (2006) Tocopherols play a crucial role in low-temperature adaptation and phloem loading in *Arabidopsis*. Plant Cell 18(10):2710–2732
Maffei ME (2010) Sites of synthesis biochemistry and functional role of plant volatiles. S Afr J Bot 76:612–631
Maffei ME, Gertsch J, Appendino G (2011) Plant volatiles: production function and pharmacology. Nat Prod Rep 28:1359–1380
Magazù S, Migliardo F, Benedetto A, La Torre R, Hennet L (2012) Bio-protective effects of homologous disaccharides on biological macromolecules. Eur Biophys J 41(4):361–367
Mahfoozi S, Limin AE, Fowler DB (2001) Developmental regulation of low-temperature tolerance in winter wheat. Ann Bot 87:751–757
Mahfoozi S, Limin AE, Ahakpaz F, Fowler DB (2006) Phenological development and expression of freezing resistance in spring and winter wheat under field conditions in north-west Iran. Field Crop Res 97(2):182–187
Mäntylä E, Lång V, Palva ET (1995) Role of abscisic acid in drought-induced freezing tolerance cold acclimation and accumulation of LTI78 and RAB18 proteins in *Arabidopsis thaliana*. Plant Physiol 107:141–148
Martin-Tanguy J (2001) Metabolism and function of polyamines in plants: recent development (new approaches). J Plant Growth Regul 34(1):135–148
Maruyama K, Sakuma Y, Kasuga M, Ito Y, Seki M, Goda H, Shimada Y, Yoshida S, Shinozaki K, Yamaguchi-Shinozaki K (2004) Identification of cold-inducible downstream genes of the *Arabidopsis* DREB1A/CBF3 transcriptional factor using two microarray systems. Plant J 38(6):982–993
Mazzucotelli E, Tartari A, Cattivelli L, Forlani G (2006) Metabolism of γ-aminobutyric acid during cold acclimation and freezing and its relationship to frost tolerance in barley and wheat. J Exp Bot 57(14):3755–3766
Megha S, Basu U, Kav NN (2014) Metabolic engineering of cold tolerance in plants. Biocatal Agric Biotechnol 3(1):88–95
Minocha R, Majumdar R, Minocha SC (2014) Polyamines and abiotic stress in plants: a complex relationship. Front Plant Sci 5:175
Morin X, Ameglio T, Ahas R, Kurz-Besson C, Lanta V, Lebourgeois F, Miglietta F, Chuine I (2007) Variation in cold hardiness and carbohydrate concentration from dormancy induction to bud burst among provenances of three European oak species. Tree Physiol 27:817–825

Morsy MR, Jouve L, Hausman JF, Hoffmann L, Stewart JM (2007) Alteration of oxidative and carbohydrate metabolism under abiotic stress in two rice (*Oryza sativa L.*) genotypes contrasting in chilling tolerance. J Plant Physiol 164(2):157–167

Nägele T, Stutz S, Hörmiller II, Heyer AG (2012) Identification of a metabolic bottleneck for cold acclimation in *Arabidopsis thaliana*. Plant J 72(1):102–114

Nagler M, Nukarinen E, Weckwerth W, Nägele T (2015) Integrative molecular profiling indicates a central role of transitory starch breakdown in establishing a stable C/N homeostasis during cold acclimation in two natural accessions of *Arabidopsis thaliana*. BMC Plant Biol 15(1):284

Nahar K, Hasanuzzaman M, Alam MM, Fujita M (2015) Exogenous spermidine alleviates low temperature injury in mung bean (*Vigna radiata L.*) seedlings by modulating ascorbate-glutathione and glyoxalase pathway. Int J Mol Sci 16(12):30117–30132

Neuhaus HE, Emes MJ (2000) Nonphotosynthetic metabolism in plastids. Annu Rev Plant Biol 51(1):111–140

Niinemets U, Loreto F, Reichstein M (2004) Physiological and physicochemical controls on foliar volatile organic compound emissions. Trends Plant Sci 9(4):180–186

Niinemets U (2010) Mild versus severe stress and BVOCs: thresholds priming and consequences. Trends Plant Sci 15:145–153

Ninagawa T, Eguchi A, Kawamura Y, Konishi T, Narumi A (2016) A study on ice crystal formation behavior at intracellular freezing of plant cells using a high-speed camera. Cryobiology 73(1):20–29

Ogasawara N, Hiramasu T, Ishiyama K, Fushimi H, Suzuki H, Takagi H (2001) Effects of gibberellic acid and temperature on growth and root carbohydrates of *Delphinium* seedlings. J Plant Growth Regul 33(3):181–187

Okumoto S, Funck D, Trovato M, Forlani G (2016) Amino acids of the glutamate family: functions beyond primary metabolism. Front Plant Sci 7:318

Park EJ, Jeknić Z, Sakamoto A, DeNoma J, Yuwansiri R, Murata N, Chen TH (2004) Genetic engineering of glycinebetaine synthesis in tomato protects seeds plants and flowers from chilling damage. Plant J 40(4):474–487

Partelli FL, Vieira HD, Rodrigues APD, Pais I, Campostrini E, Chaves MMCC, Ramalho JC (2010) Cold induced changes on sugar contents and respiratory enzyme activities in coffee genotypes. Ciência Rural 40(4):781–786

Pasandi M, Janmohammadi M (2015) Effects of cold-hardening and plant developments on freezing tolerance of winter plants in a cold climate. Curr Opin Agric 4(1):10–18

Peshev D, Vergauwen R, Moglia A, Hideg É, Van den Ende W (2013) Towards understanding vacuolar antioxidant mechanisms: a role for fructans. J Exp Bot 64(4):1025–1038

Pirzadah TB, Malik B, Rehman RU, Hakeem KR, Qureshi MI (2014) Signaling in response to cold stress. In: Plant signaling: understanding the molecular crosstalk. Springer, New Delhi, pp 193–226

Pramanik MH, Imai R (2005) Functional identification of a trehalose 6-phosphate phosphatase gene that is involved in transient induction of trehalose biosynthesis during chilling stress in rice. Plant Mol Biol 58:751–762

Prášil IT, Kadlecová-Faltusová Z, Faltus M (2001) Water and ABA content in fully expanded leaves of cold-hardened barleys. Icel Agric Sci 14:49–53

Puhakainen T (2004) Physiological and molecular analyses of cold acclimation of plants. University of Helsinki, Finland

Purdy SJ, Bussell JD, Nunn CP, Smith SM (2013) Leaves of the *Arabidopsis* maltose exporter1 mutant exhibit a metabolic profile with features of cold acclimation in the warm. PLoS One 8:e79412

Rácz I, Páldi E, Szalai G, Janda T, Pál M, Lásztity D (2008) S-methylmethionine reduces cell membrane damage in higher plants exposed to low-temperature stress. J Plant Physiol 165(14):1483–1490

Rai VK, Sharma UD (1991) Amino acids can modulate ABA induced stomatal closure stomatal resistance and K+ fluxes in Vicia faba leaves. Beitr Biol Pflanz 66:393–405

Rai VK (2002) Role of amino acids in plant responses to stresses. Biol Plant 45(4):481–487

Rao RSP, Andersen JR, Dionisio G, Boelt B (2011) Fructan accumulation and transcription of candidate genes during cold acclimation in three varieties of *Poa pratensis*. J Plant Physiol 168(4):344–351

Rekarte-Cowie I, Ebshish OS, Mohamed KS, Pearce RS (2008) Sucrose helps regulate cold acclimation of *Arabidopsis thaliana*. J Exp Bot 59(15):4205–4217

Rosa M, Prado C, Podazza G, Interdonato R, González JA, Hilal M, Prado FE (2009) Soluble sugars: metabolism sensing and abiotic stress: a complex network in the life of plants. Plant Signal Behav 4(5):388–393

Ruan YL (2014) Sucrose metabolism: gateway to diverse carbon use and sugar signaling. Annu Rev Plant Biol 65:33–67

Ryu SB, Costa A, Xin Z, Li PH (1995) Induction of cold hardiness by salt stress involves synthesis of cold- and abscisic acid-responsive proteins in potato (*Solanum Commersonii Dun*). Plant Cell Physiol 36:1245–1251

Sah SK, Kaur G, Wani SH (2016) Metabolic engineering of compatible solute trehalose for abiotic stress tolerance in plants. In: Osmolytes and plants acclimation to changing environment: emerging Omics technologies. Springer, New Delhi, pp 83–96

Sakamoto A, Murata N (2002) The role of glycine betaine in the protection of plants from stress: clues from transgenic plants. Plant Cell Environ 25(2):163–171

Samanta A, Das G, Das SK (2011) Roles of flavonoids in plants. Int J Pharm Sci Technol 6:12–35

Savitch LV, Gray GR, Huner NP (1997) Feedback-limited photosynthesis and regulation of sucrose-starch accumulation during cold acclimation and low-temperature stress in a spring and winter wheat. Planta 201(1):18–26

Schneider T, Keller F (2009) Raffinose in chloroplasts is synthesized in the cytosol and transported across the chloroplast envelope. Plant Cell Physiol 50(12):2174–2182

Schulz E, Tohge T, Zuther E, Fernie AR, Hincha DK (2016) Flavonoids are determinants of freezing tolerance and cold acclimation in *Arabidopsis thaliana*. Sci Rep 6:34027

Sharkey TD, Yeh S (2001) Isoprene emission from plants. Annu Rev Plant Physiol Plant Mol Biol 52:407–436

Sharma N, Cram D, Huebert T, Zhou N, Parkin IA (2007) Exploiting the wild crucifer Thlaspi arvense to identify conserved and novel genes expressed during a plant's response to cold stress. Plant Mol Biol 63(2):171–184

Shen B, Hohmann S, Jensen RG, Bohnert HJ (1999) Roles of sugar alcohols in osmotic stress adaptation. Replacement of glycerol by mannitol and sorbitol in yeast. Plant Physiol 121(1):45–52

Shi J, Fu XZ, Peng T, Huang XS, Fan QJ, Liu JH (2010) Spermine pretreatment confers dehydration tolerance of citrus in vitro plants via modulation of antioxidative capacity and stomatal response. Tree Physiol 30(7):914–922

Shirasawa K, Takabe T, Takabe T, Kishitani S (2006) Accumulation of glycinebetaine in rice plants that overexpress choline monooxygenase from spinach and evaluation of their tolerance to abiotic stress. Ann Bot 98(3):565–571

Singh BD (2014) Plant breeding: principles and methods. Kalyani Publishers, New Delhi

Slewinski TL (2012) Non-structural carbohydrate partitioning in grass stems: a target to increase yield stability, stress tolerance, and biofuel production. J Exp Bot 63(13):4647–4670

Smallwood M, Bowles DJ (2002) Plants in a cold climate. Philos Trans R Soc Lond Ser B Biol Sci 357:831–847

Solanke AU, Sharma AK (2008) Signal transduction during cold stress in plants. Physiol Mol Biol Plants 14(1–2):69–79

Stefanowska M, Kuraś M, Kacperska A (2002) Low temperature-induced modifications in cell ultrastructure and localization of phenolics in winter oilseed rape (*Brassica napus L. var. oleifera L.*) leaves. Ann Bot 90(5):637–645

Strimbeck GR, Schaberg PG, Fossdal CG, Schröder WP, Kjellsen TD (2015) Extreme low temperature tolerance in woody plants. Front Plant Sci 6:884

Taji T, Ohsumi C, Iuchi S, Seki M, Kasuga M, Kobayashi M, Yamaguchi-Shinozaki K, Shinozaki K (2002) Important roles of drought- and cold inducible genes for galactinol synthase in stress tolerance in *Arabidopsis thaliana*. Plant J 29:417–426

Takahashi T, Kakehi JI (2010) Polyamines: ubiquitous polycations with unique roles in growth and stress responses. Ann Bot 105(1):1–6
Tapernoux-Lüthi EM, Böhm A, Keller F (2004) Cloning functional expression and characterization of the raffinose oligosaccharide chain elongation enzyme galactan: galactan galactosyltransferase from common bugle leaves. Plant Physiol 134(4):1377–1387
Tarkowski ŁP, Van den Ende W (2015) Cold tolerance triggered by soluble sugars: a multifaceted countermeasure. Front Plant Sci 6:203
Theocharis A, Clément C, Barka EA (2012) Physiological and molecular changes in plants grown at low temperatures. Planta 235(6):1091–1105
Thomashow MF (1999) Plant cold acclimation: freezing tolerance genes and regulatory mechanisms. Annu Rev Plant Physiol Plant Mol Biol 50:571–599
Turhan E, Ergin S (2012) Soluble sugars and sucrose-metabolizing enzymes related to cold acclimation of sweet cherry cultivars grafted on different rootstocks. Sci World J 2012 Article 7 pages
Tuteja N, Mahajan S (2007) Calcium signaling network in plants: an overview. Plant Signal Behav 2(2):79–85
Valle RVB (2002) Mechanisms of frost adaptation and freeze damage in grapevine buds. PhD dissertation Hohenheim University
Van Velthuizen H (2007) Mapping biophysical factors that influence agricultural production and rural vulnerability (No. 11). Food & Agriculture Organization
Verbruggen N, Hermans C (2008) Proline accumulation in plants: a review. Amino Acids 35(4):753–759
Vickers CE, Gershenzon J, Lerdau MT, Loreto F (2009) A unified mechanism of action for volatile isoprenoids in plant abiotic stress. Nat Chem Biol 5:283–291
Wani SH, Brajendra Singh N, Haribhushan A, Iqbal Mir J (2013) Compatible solute engineering in plants for abiotic stress tolerance-role of glycine betaine. Curr Genomics 14(3):157–165
Wani SH, Sah SK, Hossain MA, Kumar V, Balachandran SM (2016) Transgenic approaches for abiotic stress tolerance in crop plants. In: Advances in plant breeding strategies: agronomic abiotic and biotic stress traits. Springer, Cham, pp 345–396
Xin Z, Browse J (2000) Cold comfort farm: the acclimation of plants to freezing temperatures. Plant Cell Environ 23(9):893–902
Xing W, Rajashekar CB (2001) Glycine betaine involvement in freezing tolerance and water stress in *Arabidopsis thaliana*. Environ Exp Bot 46(1):21–28
Yadav SK (2010) Cold stress tolerance mechanisms in plants. A review. Agron Sustain Dev 30(3):515–527
Yue C, Cao HL, Wang L, Zhou YH, Huang YT, Hao XY, Wang YC, Wang B, Yang YJ, Wang XC (2015) Effects of cold acclimation on sugar metabolism and sugar-related gene expression in tea plant during the winter season. Plant Mol Biol 88(6):591–608
Yuki M, Grukhin V, Lee CS, Haworth IS (1996) Spermine binding to GC-rich DNA: experimental and theoretical studies. Arch Biochem Biophys 325(1):39–46
Zhang J, Wu X, Niu R, Liu Y, Liu N, Xu W, Wang Y (2012) Cold-resistance evaluation in 25 wild grape species. Vitis 51(4):153–160
Zhang W, Jiang B, Li W, Song H, Yu Y, Chen J (2009) Polyamines enhance chilling tolerance of cucumber (*Cucumis sativus L.)* through modulating antioxidative system. Sci Hortic 122(2):200–208
Zhao H, Yang H (2008) Exogenous polyamines alleviate the lipid peroxidation induced by cadmium chloride stress in *Malus hupehensis* Rehd. Sci Hortic 116:442–447

[illegible]

Chapter 9
Breeding Cold-Tolerant Crops

Elisabetta Frascaroli

9.1 Introduction

Low-temperature stress is considered as the major abiotic constraint limiting plants' growth and the exploitation of the potential land cultivation (Jha et al. 2017). Crops evolved, through domestication, generally from warm areas of the planet (Feldman and Levy 2009), thanks to acclimation and adaptation mechanisms that made it possible for them to survive at higher latitudes. As adaptation to cold was undertaken by most of the crop plants, the traits that allow a satisfactory yield are shared among different species and present functional conservation among them. These common tolerance traits can be the effect of *convergent evolution* via independent paths to a similar outcome or of *monophyletic origin* if they all descend from a common ancestor (Mickelbart et al. 2015). Indeed, several specific adaptations are involved in the maximal exploitation of yield potential of a crop (Huner et al. 2014). Among them, we name the modulation of the amount of energy that can reach the plant, the light interception efficiency, the energy conversion efficiency, and the partitioning efficiency of products into commercial yield. All those steps can be limited by less than optimal temperature. In addition, to adapt to new, cooler environments, plants need to modify a number of traits, like photoperiod sensitivity, as in case of wheat (Guo et al. 2018), maize, and potato, or vernalization, e.g., for wheat, barley, and carrots (Mickelbart et al. 2015).

Crop adaptation to limiting low temperature still is an important breeding objective, as observed by Bradshaw (2017) who also noticed that "breeders still need to

E. Frascaroli (✉)
Department of Agricultural and Food Sciences (DISTAL), University of Bologna, Bologna, Italy
e-mail: elisabetta.frascaroli@unibo.it

S. H. Wani, V. Herath (eds.), *Cold Tolerance in Plants*,
https://doi.org/10.1007/978-3-030-01415-5_9

apply appropriate breeding methods to the right germplasm for the right objectives". The objective of breeding is determined by the characteristics of the new and more efficient farming systems and of the innovative uses of crops, in turn devised to face the increased need of food and energy expected in the near future. When environmental conditions deviate from the optimal temperature, growth is limited, and there is risk of tissue damage or even of impaired survival ability (Korner 2016). In general, conditions with temperatures lower than optimum can be found at high elevation or at high latitude, but there are cold periods in temperate zones, too. Genotypes able to survive in extreme environments, like in case of alpine trees (Korner et al. 2016), are able to actively grow during the period with acceptable temperature. Plants can adapt to temperatures that become low only seasonally by means of modification of phenology and morphology. In higher latitudes, plants can basically develop only by using the warmest periods, and thus they need to grow faster to take advantage of the otherwise short season. Morphological modifications concern usually plant size, in that smaller plants are generally less susceptible than taller ones.

Abiotic stresses impose negative deviations from the maximum yield potential, and thus they are important factors determining crop yield stability (Mickelbart et al. 2015). Traditionally, increased yield stability has been attained through selective breeding, and selection has been successful in changing allelic frequencies of favorable alleles, although taking a long time to accomplish the change. However, since abiotic stresses, including cold stress, are somehow unpredictable in the field, sometimes it is more advisable to select genotypes with high potential *per se* and wide adaptability and stability, instead of trying to develop varieties designed for a narrow agroecological region and a specific stress (Arief et al. 2015; Bennett et al. 2012; Stojakovic et al. 2015).

In recent decades, crops had to face changing environmental conditions, with a general increase in the mean temperature and especially a higher temperature fluctuation (Xu et al. 2018). Even though, at first glance, the global warming might seem to reduce frost damage in crops (Cattivelli 2011), the fluctuation of winter temperatures, with warm periods, has a strong impact on the frost tolerance both in crops and natural plant species. In fact, warm days during the winter can strongly interfere with dormancy and vernalization, so plants start an active growth and lose most of their freezing tolerance. In addition, with warmer winters, temperate tree species tend to show earlier flowering, with increased risk of damage during the spring frost (Eccel et al. 2009). Modification in the cultivation practices to cope with new environmental conditions have already been observed, with significant changes in tillage techniques and in the timing of cultivation toward earlier sowing, to avoid dry periods during the summer and to use as much as possible of the winter and spring precipitations (Olesen et al. 2011, 2012; Trnka et al. 2011). Indeed, temperature changes will trigger the need for optimization of the crop growth phases, to minimize the stress and to maximize the productivity of the agronomic resources employed. New tillage practices are focused on soil–water conservation and protection against soil erosion, as these issues are believed to become increasingly important (Falloon and Betts 2010). These practices, however, leave the soil colder than the traditional ones for the sowing of spring crops, and thus genotypes tolerant to low temperature at germination will better suit the new practices. As for the timing of cultivation, a study regarding the major trends in Europe (Olesen et al. 2012) pointed out that "climate change will affect the timing of cereal crop development, but exact changes will also depend on changes in varieties as affected by plant breeding and variety choices". The modification of the timing of crop growth and of the agronomic practices can be made possible through plant breeding by selecting for different traits, such as the response to photoperiod or the response to temperature critical for each growth phase. Genotypes selected for tolerance to suboptimal temperature are also particularly useful when cultivation is addressed to the low input agriculture or for high elevated lands (Sthapit and Witcombe 1998). This is the case of both the low input climate-smart agriculture and the highly efficient sustainable intensification (Campbell et al. 2014; Garnett et al. 2013; Nemali et al. 2015; Thakur and Uphoff 2017). Agronomic models have been developed to describe the growth phases of major crops under different climate conditions (Waha et al. 2012) and for the characterization of breeding needs in more adapted varieties through the identification of an appropriate ideotype (Olesen et al. 2012; Semenov et al. 2014; Semenov 2009).

This chapter will review (1) how cold tolerance can be the key of the crop adaptation to low input and highly efficient agricultural techniques and (2) how plant breeding can integrate classical and molecular methods to select genotypes maximizing crop performance in cold environments.

9.2 Cold Tolerance in Adaptation to Environment

Selection of alleles for adaptation to favorable and unfavorable environments already led to a general improvement of crops yield. This achievement was accomplished through effective stress adaptation, namely, to low and subfreezing temperatures, high temperature, flooding, drought, salinity, ion toxicity, ion deficiency, and ozone (Mickelbart et al. 2015). Adaptation of plants to abiotic stresses, characterizing diverse and variable environments, involves a variety of plasticity mechanisms and provides plants with increased yield stability. In case the suboptimal condition occurs when the plant is growing slowly, the effect may be not as great as if it occurs during plant fast growing (Dolferus 2014; Mansouri-Far et al. 2010). Moreover, stresses often occur in combinations (i.e., high temperature and drought, low temperature and flooding), so yield adaptation usually implies enhancement of multiple-stress tolerances. Oftentimes, constraints imposed by one stress are the same as another. With reference to cold, it can involve membrane damage as in case of high temperature (Bita and Gerats 2013; Dhillon et al. 2010), so the maintenance of membrane function is crucial to achieve the tolerance to both. Accordingly, a large overlapping among adaptive responses to cold and drought have been reported (Hussain et al. 2018). As an example, maintenance of water potential is a factor enhancing tolerance to cold, and such a mechanism is shared with other stresses such as drought and salinity. The response to all those environmental factors has been studied and proved to involve cross talk between various stress signaling metabolic pathways (Cramer et al. 2011; Nakashima et al. 2014; Seki et al. 2003). Tolerance or susceptibility to these stresses are complex traits, as stress may affect multiple stages of plant development, and often several stresses concurrently affect the plants (Chinnusamy et al. 2004; Li et al. 2014). Moreover, it has been reported in a number of studies that the same genes can be responsible for reaction to multiple stresses (Bai et al. 2018; Banerjee and Roychoudhury 2018; Dubois et al. 2018; Hossain et al. 2018; Mustafavi et al. 2018).

9.2.1 Cold Limits Crop Growth in Less than Optimal Environments

The ability of crops to grow in different environment/climates is one of more powerful tools enabling farmer to satisfy the growing demand of food and the changing needs of industrial crops. Plant breeding is a key process in the achievement of plants adapted to different climates (Atlin et al. 2017). Breeding for resilient crops, therefore, is one of the major strategies to cope with the increasing challenges in world agriculture (Obata et al. 2015). It is important to note that abiotic stresses inherent with the environment adversely interfere with growth and with agronomic performance through modifications regarding morphological, physiological, and molecular adjustments (Sanghera et al. 2011). Different environmental constraint exacerbates the need of cropping systems continually updated. As already noted, although it may seem counterintuitive, even the effects of increase of temperature can be overcome by means of cold tolerance. In fact, one way to cope with high temperatures and shortage of water is either to change the sowing date or to move to higher latitudes. Moreover, the rise of temperatures can affect cold acclimation and thus even impair plant overwintering survival (Arora 2018; Rapacz et al. 2014). Widespread efforts to identify major genes controlling tolerance to stress lead to the identification of some alleles with large effects for tolerance (Thoen et al. 2017). On the contrary, traits controlling stress responses are more often controlled by many genes with small effects (Haak et al. 2017).

Understanding tolerance to stress should take into account "deciphering the environmental impact on plants" (Bloomfield et al. 2014; Xu 2016). With the new appealing term envirotyping, the fine study of the genotype-by-environment interaction (GEI) has been proposed as a tool useful to the modeling of crop and to phenotype prediction (Xu 2016). The characterization of the crop target environment can be seen as a third "typing" technology, complementing with genotyping and phenotyping to study abiotic stress and to accomplish selection. Environmental information can be collected through multiple environmental trials, geographic and soil information systems, measurement of soil and canopy properties, and evalua-

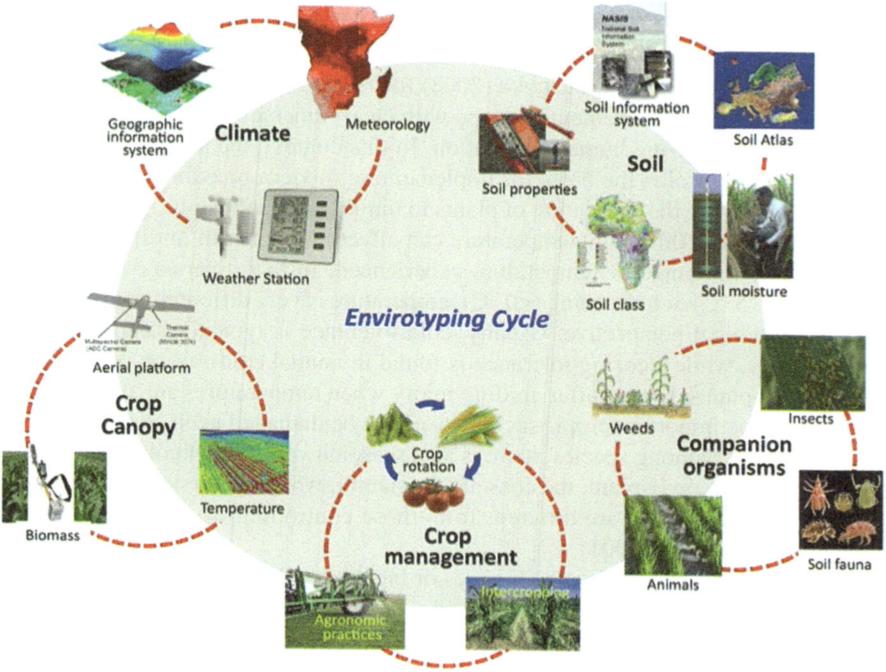

Fig. 9.1 Envirotyping. Advances in the area of integrate analysis of environment profiling contributes to crop modeling and phenotype prediction through its functional components. Envirotyping is the result of integration of environmental data by means of information and support systems. Factors to be considered include geographical and meteorological data, soil properties, data from living organisms characterizing the environment, crop management, and status of the crop, each containing several subgroups describing factors affecting plant growth and development (Xu 2016).

tion of companion organisms. All the elements of envirotyping are represented in Fig. 9.1. Envirotyping contributes to crop modeling and phenotype prediction through its functional components, including GEI (Cooper et al. 2014). Envirotyping is driven by information and support systems; it has a wide range of applications, including environmental characterization, GEI analysis, phenotype prediction, near-iso-environment construction, agronomic genomics, precision agriculture and breeding. Envirotyping contributes to the development of a four-dimensional profile of the crops, involving genotype, phenotype, envirotype, and time, considered as developmental stage (Xu 2016). The reaction of a trait to different environments, or reaction norm, is also referred as phenotypic plasticity, and its genotypic variation can parallel GEI, as reviewed by Marais et al. (2013). Advances in the area of integrate analysis of environment profiling already led to the development of environmental indexes, to be integrated in comprehensive crop models (Li et al. 2018a).

9.2.2 Genetic Control of Tolerance to Low Temperature

The review of Takeda and Matsuoka (2008) focused on the promising genetic tools available for crop improvements to cope with environmental stress and to maintain yield to feed a growing human population. In this context, the study of cold tolerance genetic control is the basis for implementing modern breeding programs and thus for improving the adaptation of plants to limiting environments. As pointed out by Revilla et al. (2005), low temperature can affect plant growth in different ways, according to the range of temperature experienced. In fact, tolerance to cold and chilling (0–15 °C) or to freezing (<0 °C) temperature is very different from a genetic and physiological perspective. Usually, cold tolerance is typical of annual warm-season crops, while freezing tolerance is found in annual cold-season or perennial crops. Most plants do not suffer chilling injury when temperatures are above 10 °C, although some important crops, such as rice, can be damaged even at 15 °C. Other than differences among species, there is also seasonal variation of cold tolerance as in the case of a woody plant, the coastal Douglas-fir, where factors controlling frost tolerance in the autumn are different from those controlling frost tolerance in the spring (Jermstad et al. 2001).

It is not completely clear which genes or biochemical processes are essential to the achievement of freezing tolerance and which ones affect responses to the low, non-freezing, temperatures but are not involved in freezing tolerance. For example, the loci involved in signal cascades mediating most aspects of cold acclimation, such as increases in abscisic acid, synthesis of compatible osmolytes, and changes in membrane lipid composition, are mostly unknown, besides for the induction of some *COR* genes (Xin and Browse 2000). Plants respond and adapt to survive under cold stress conditions by adjusting at the molecular and cellular levels, as well as at the physiological and biochemical levels (Sanghera et al. 2011). However, complexity of the response in terms of gene expression has been demonstrated to be widely variable in different genetic backgrounds. In other words, the response to the stress is different depending on the genotype, because the factor limiting tolerance can be different in each background (Canas et al. 2017). The role of phenotypic plasticity in adaptation to nonoptimal environments and its genetic control has been debated for long time (Josephs 2018; Marais et al. 2013; Via et al. 1995). Estimates of phenotypic plasticity were used to identify loci associated with GEI or to stress tolerance. The results obtained so far indicate that traits per se and plasticity may be controlled by different sets of genes (Kusmec et al. 2017). On the other hand, candidate genes for stress tolerance, like those involved in phytohormone-mediated processes, were proved to be involved in multiple stress responses (Kusmec et al. 2017). Some of them showed contrasting effects with different stresses (adaptive genes), while other showed consistent effects across different stress conditions (constitutive genes) (Thoen et al. 2017).

Several studies suggest that in some cases cold tolerance may be under quite simple genetic control. Stone et al. (1993) found that freezing tolerance and acclimation capacity are under control of relatively few genes in populations derived from interspecific crosses between *Solanum commersonii* and *Solanum cardiophyllum*. Extensive examples of cold tolerance genetic control is reported by Revilla et al. (2005), who pointed out that tolerance at different growth stages are definitely under the control of different gene sets. Since low temperature resistance in plants seems to be most of the times a very complex trait, involving many different metabolic pathways and cell compartments, physiological breeding has been suggested as a possible way to bring together different components of tolerance, starting with crosses of genotypes with complementary traits (Reynolds and Langridge 2016). To assess the predictive value of those physiological traits for yield and its components, correlation studies and network analyses have been performed (Obata et al. 2015).

9.3 Cold Tolerance in Plant Breeding

As already mentioned in the previous chapters, plant breeding is the key process for the achievement of plants adapted to different climates. Conventional breeding methods have met limited success in improving the cold tolerance of important crop plants based on interspecific or intergeneric hybridization (Jha et al. 2017). The conventional breeding approaches are limited by the complexity of stress tolerance traits, low genetic variance of yield, low heritability of yield components under stress conditions, and lack of efficient selection criteria (Sanghera et al. 2011). However, ample genetic reservoir for cold tolerance can be available in well-adapted breeding populations. Moreover, germplasm collected from high-altitude and low-temperature areas, cold-tolerant mutant, somaclonal variants, and wild species can be exploited for breeding improved cold-tolerant genotypes (Sanghera et al. 2011).

9.3.1 Genetic Variability

In order to successfully apply any breeding program, what is needed first is a suitable genetic diversity, but it is also crucial to develop a deep understanding of the trait(s) to select for. So, according to Dwivedi et al. (2017), assessing the proper functional diversity determining yield in less than optimal environment implies to decide what are the targets of selection and what are the best methods to use. As the first step is to find genetic variability, genetic resources can be considered to access favorable alleles available in the germplasm collections. As pointed out by Revilla et al. (2005), genes for cold tolerance can be transferred from a source to a recipient genotype more easily in case the genetic distance between the two is not very high. The most common way to include a donor genotype in a breeding program is to cross a tolerant accession and a susceptible élite genotype and to select within the segregating population. For example, Zhang et al. (2014) were able to obtain a significant response to selection for cold tolerance in rice from crosses of susceptible by resistant parents. Recurrent selection programs (Frascaroli and Landi 2013;

Zhang et al. 2014) already obtained appreciable responses in the segregating populations developed from a cross between two inbreds of different origin, one elite and one locally adapted, both with an average tolerance to cold. In that case, the response to selection was due to the recovery of transgressive individuals originating from recombination. In other cases, favorable alleles were transferred within the same species, from one maize type (i.e., field maize) to another specialty maize (i.e., sweet maize) (Revilla et al. 1998) or from winter to spring wheat (Braun et al. 1996). Exploitation of natural variation for tolerance to abiotic stresses can also rely on the use of landraces (Dwivedi et al. 2013, 2016).

To improve the chances of a successful breeding, stress adaptation loci involved in yield stability and in field performances under environmental extremes were investigated. In particular, for quantitative traits, quantitative traits loci (QTL) involved in the control of adaptation to cold can be identified through linkage mapping based on a mapping population or through genome-wide association studies (GWAS) on genetic panels with genotypes of different origin. According to Thoen et al. (2017), investigation on plant QTL controlling tolerance to stress brings some inherent difficulties due to the complexity of the response. In fact, different stresses are often present simultaneously, as in case of cold and drought, anoxia, and so on. Moreover, phenotype expression in response to two biotic stresses could not be predicted on the basis of existing information regarding interactions between underlying signaling pathways. In addition, cross talk among the responses to various stress complicate further the picture. In spite of those limitations, the study and the dissection of the genetic control of cold tolerance have been pursued for different species. Several studies on QTL analysis for cold tolerance have been already reported by Revilla et al. (2005). In more recent days, thanks to the advancements of the knowledge and to the lowering of genotyping costs, studies gained higher impact, counting on the size of mapping populations. A wide review of the literature is reported by Jha et al. (2017). For barley, QTL analysis in the "Nure" (winter) × "Tremois" (spring) cross (Francia et al. 2004) and the fine mapping on more than 1800 recombinants was carried out (Francia et al. 2007) for the two major low-temperature tolerance QTL, i.e., *Fr-H1* and *Fr-H2*. In case of winter wheat, Zhao et al. (2013) analyzed a large mapping population of 1739 genotypes. In maize some work was focused on the variation in response to cold for chlorophyll content or photosynthesis (Fracheboud et al. 2004; Hund et al. 2005; Rodriguez et al. 2014; Strigens et al. 2013). Given the complexity of the trait, QTL mapping must take advantage of large and complex populations. In maize, for example, 720 double haploids were investigated for adaptation to chilling conditions to map genome regions involved in tolerance (Presterl et al. 2007). In rice almost 2000 lines were considered for fine mapping (Andaya and Mackill 2003). Complex mapping populations, like connected populations, have been used for mapping QTL for tolerance to cold, as in case of maize at germination phase (Li et al. 2018b) where 650 families allowed to map up to 43 QTL, that reduced to three after a meta-analysis of the three connected populations.

New techniques aimed at the identification of specific genes for cold tolerance can be pursued by using transcriptomic and proteomic approaches and/or QTL validation and cloning (Marla et al. 2017; Salvi and Tuberosa 2015; Sheng et al. 2017). A rich analysis of the plethora of results obtained with those methods can be found in dedicated chapters. However, an important effort is needed to integrate all genomics information of crop and model species into databases to make it possible the comparative analysis of genomes, also in response to stress conditions (Naithani et al. 2017; Tello-Ruiz et al. 2018).

9.3.2 Breeding Programs

The objective of breeding is to obtain new varieties that are improved for their tolerance or resistance to different stresses (Trachsel et al. 2017).The definition of selection criteria is an essential step in any breeding program and particularly so when trying to improve cold tolerance (Paleari et al. 2017). In turn, to choose those criteria, plant breeders made effort to model and understand GEI (Crespo-Herrera et al. 2017; Lado et al. 2016) in order to well define mega-environments, in the attempt to select for the tolerance to the more frequent stresses limiting adaptation to a particular region.

Successful strategies must be devised to improve the élite varieties, depending on the genetic control of the trait. In case tolerance is due to just one locus, the good allele, usually found in a non-élite genotype, can be introgressed into the élite one. Introgression is carried out by repeated backcrosses aimed at transferring a genetic determinant (allele) from the donor to the recipient genotype. Introgression can be accelerated by the use of molecular markers for the donor chromosomal region, as well as of markers surrounding that region and the other chromosomes of the recipient genotype. Altogether, backcross breeding is the most common strategy to improve single-target traits, particularly the high heritable ones (Bernardo 2016b). This method proved to be successful to improve yield and cold tolerance together in rice (Meng et al. 2013; Zhang et al. 2014) and to select for cold tolerance even in case of complex genetic control (Zhang et al. 2014).

Breeding for stress tolerance or avoidance, and especially for cold tolerance, has proved to be challenging, at least partly because tolerance mechanisms are often environment-specific, and screening methods that integrate the multiple spatial and temporal variations relevant to this stress are difficult to establish. For the need to adapt to all these multiple spatial and temporal variations characterizing cold environment, tolerance is often controlled by many loci, whose small effects combined confer tolerance to stress. This complex control implies that breeding for tolerant genotypes cannot be pursued as for the monogenic traits. Moreover, the small effect of each quantitative locus is usually subjected to a relevant genotype-by-environment interaction, thus reducing the effectiveness of multiple loci detection when analyzed in multiple environments (Makumburage et al. 2013). Selection can be effective both for high and low tolerance, as in case of Landi et al. (1992) who obtained significant responses in divergent full-sib recurrent selection for the difference between germination at low and at optimal temperature, i.e., for the reaction norm, in maize.

Response to selection was appreciable in cold conditions both in controlled environment and in the field (Frascaroli and Landi 2013) but not when the selected genotypes were compared at warm temperature. Later studies evaluated the same populations focusing on associated changes in mitochondrial properties and found a correlated response concerning the interaction between membrane lipids and cytochrome-c-oxidase content (De Santis et al. 1999; Tampieri et al. 2011).

Breeding for cold tolerance can be limited by major bias due to environmental conditions during the evaluation of the trait. In fact, field-breeding programs suffer environmental variations that limit selection progress (Ly et al. 2018), while programs carried out under controlled conditions in the laboratory are limited by the correlation between field and laboratory performance. An accurate, consistent technique would therefore be very helpful. There are examples where reliable methods or indexes can be used as indicators of tolerance in crops under selection (Thapa et al. 2008). Cold chambers are often used for selecting for cold tolerance and for integrating evaluations made in the field (Frascaroli and Landi 2013; Revilla et al. 2014, 2016). The evaluation of reaction to cold during selection can also be impaired by large experimental errors, often a side effect of the stressful condition itself, in addition to the experimental error due to the methodologies and the interaction with unpredictable environmental variation. To overcome this drawback, high-throughput phenotyping systems can be considered, once the experimental procedures are adequately optimized (Junker et al. 2015; Tschiersch et al. 2017).

There are not many recent published reports on the results obtained with the most common breeding methods, probably because these applied methods are designed for improving varieties and not for publishing scientific articles, owing to the lack of systematic design or the absence of novelty. Moreover, uncontrolled environmental variation in breeding programs can limit genetic gains for cold tolerance achievable through selection. Indeed, as previously explained, selection entails some problems of evaluation because of the unpredictable climatic variation of field trials and the inconsistent correlation between controlled environment and field performance. However, breeding designed to adapt maize to Northern Europe has resulted in the release of compact maize hybrids (Frei 2000). Similarly, in Canada most of the genetic yield improvement of maize is attributable to increased cold stress tolerance (Assefa et al. 2017; Tollenaar and Wu 1999). In tomato, Foolad and Lin (2001) obtained with mass selection significant improvements in germination under cold conditions. On the contrary, according to Hensleigh et al. (1992), conventional breeding methods have not been successful in developing barley cultivars with adequate winter hardiness for many northern regions.

9.4 Prospective for Breeding

Selection has been an important human activity since the domestication of crops, but after the advent of the genetic studies, the application of statistics and of the scientific method, breeding has developed its own theory, and selection programs can be planned and developed successfully. The more recent ability to dissect tolerance loci by means of molecular genetics gave rise to a burst of speed in the understanding of the mechanism of tolerance, even though this knowledge is not immediately translatable into a useful breeding procedure when many loci are involved. The availability of élite cultivars and the ability to perform rapid breeding cycles, providing the farmer with new cultivars for specific conditions, are important elements to adapt the cropping system to a new environment or agronomic technique (Atlin et al. 2017). To be ready and flexible for new environmental constraints, breeding programs should evaluate the potential of new cultivars in a wide range of climatic conditions. To reach this goal, according to the plant breeding theory, the main steps are (i) the improvement of base populations, (ii) the selection of commercial élite varieties, and (iii) the dissemination of the new varieties. Recurrent selection is the breeding method of choice to enrich a population with favorable alleles for a polygenic trait (Bernardo 2010; Gorjanc et al. 2018; Mueller et al. 2018). This method has been successfully employed to improve breeding population for stress tolerance, as reported, for example, by Meng et al. (2013) who was able to obtain a large numbers of cold-tolerant lines in rice. Successful breeding programs were carried out in maize (Frascaroli and Landi 2016; Sezegen and Carena 2009; Viesselmann et al. 2014), chickpea (Jha et al. 2017), ryegrass (Iraba et al. 2013), and alfalfa (Castonguay et al. 2009). The selected genotype can also be considered useful materials for genetic, physiological, and molecular dissection of cold tolerance traits using DNA markers and other -omic tools (Frascaroli and Landi 2018; Meng et al. 2013) to obtain indication on the genetic control of the trait and the genomic regions involved in the response to selection.

It has been observed that in most of the cases, the genetic improvement currently possible with conventional approaches does not exceed the 1% per year (Fischer and Edmeades 2010). For this reason different strategies for improving the selection efficiency were explored. As observed by Mickelbart et al. (2015), a combination of different approaches is advisable to accelerate the identification and characterization of specific loci that can be moved by molecular marker-assisted selection (MAS) into elite varieties, minimizing yield-adverse linkages. According to Reynolds and Langridge (2016), physiological breeding can be the strategic tool to speed up the improvement in crop adaptation and, ultimately, the ability to provide food and other products needed for the world improving population. As already mentioned, the choice of the more appropriate method must be made according to the genetic control of the traits. Indeed, selection approaches are different for traits at different levels of integration and genetic complexity, as shown in Fig. 9.2, from Reynolds and Langridge (2016). The figure represents plant selection approaches that may be used for traits controlled at different levels of complexity, starting with simple metabolites and culminating in polygenic productivity traits such as yield and biomass. Current research is expected to yield (i) basic insights on cold tolerance, with advanced information on the affected molecular and physiological processes, and (ii) applied tools such as the identification and the characterization of useful genes for improving tolerance to cold stress. This information could be useful to obtain allele sequences of functionally characterized genes from which

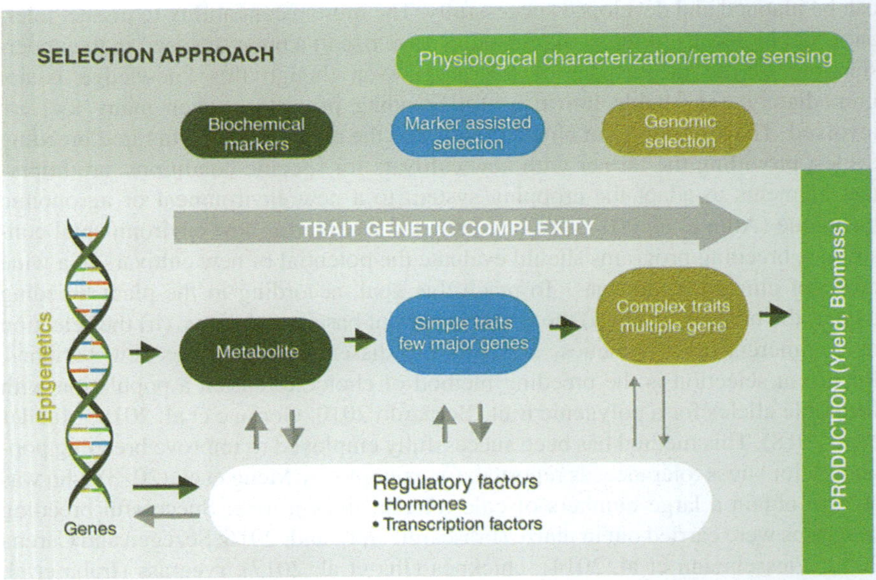

Fig. 9.2 Traits with different levels of integration and genetic complexity are selected with different approaches. Traits represented by simple metabolites can be followed; thanks to biochemical markers, simple traits controlled by major genes are conveniently selected by means of marker-assisted selection (MAS), while selection for complex traits, such as most of the cold tolerance and productivity traits, must rely on comprehensive methods like genomic selection (GS). The whole picture takes into account also regulatory factors that may interact with the expression of traits at any level as they interact with the environment (epigenetics) and thus enhance the genotype-by-environment interaction (GEI). (From Reynolds and Langridge (2016))

functional motifs affecting plant phenotype can be identified and may be used as functional markers (Andersen and Lubberstedt 2003; Brenner et al. 2013). However, it has also been observed that QTL mapping and its application in breeding are most useful for traits, for example, wheat Fusarium head blight resistance or soybean cyst nematode resistance, which might have one or a few underlying major loci (Bernardo 2008).

In spite of the huge effort made to understand all the components determining a trait, the consensus among plant breeders today is still that it is usually more efficient to select for the primary trait itself, rather than to select for multiple secondary traits that are components of or associated with it (Bernardo 2016a). As for the use of single genes, so far, the transgenic approach has not been able to handle polygenic traits, while it may be usefully utilized in cultivars transformed with a single transgene (Bernardo 2016a). In the recent years, we witnessed the enhancement of knowledge of plant genomics, and especially of sequence technology, and the shifting of plant science from "explanatory" to "predictive." Indeed, the possibility to predict the optimal genotype based on genomic information would greatly enhance the efficiency of plant breeding programs. Dense genotyping, that provided a very large amount of SNP (Single Nucleotide Polymorphism) markers, made it possible to develop new promising methods for improving complex traits that are controlled by many QTL with small effects (Bernardo 2016a). Genomic selection (GS) is a breeding method that accelerate the selection of genotypes carrying favorable alleles at loci undetectable by means of mapping models (Meuwissen et al. 2001) and can be integrated into recurrent selection (Gorjanc et al. 2018; Mueller et al. 2018). Genome-wide selection is expected to be particularly efficient when phenotypic selection is nonexistent or ineffective. Simulation and empirical studies revealed that GS can even be more efficient than MAS (Bernardo and Yu 2007), especially for traits that are difficult to measure such as cold tolerance, particularly in the field. The most advanced prediction techniques (Crossa et al. 2017; Heffner et al. 2010; Montesinos-Lopez et al. 2018) are promising in the perspective of selection implementation for cold tolerance in combination with other agronomic traits.

In conclusion, crop adaptation is needed to increase production and stability under cold conditions that are getting worse with climatic change. To improve crop plants for complex traits, such as cold tolerance, the key will be the combination of classical plant breeding with the advances in genomics, crop physiology, and modeling in an integrated profile involving genotype, phenotype, and envirotype. In particular, the most promising approach for selection will involve high-precision, high-throughput phenotyping in controlled growth chambers and platforms, and multi-environment field trials combining agronomic, morphological, physiological, and biochemical data. The exploitation of big data will involve updated statistical models for mapping genomic regions controlling cold tolerance or for predicting the breeding value of genotypes.

References

Andaya VC, Mackill DJ (2003) Mapping of QTLs associated with cold tolerance during the vegetative stage in rice. J Exp Bot 54(392):2579–2585

Andersen JR, Lübberstedt T (2003) Functional markers in plants. Trends in Plant Science 8 (11):554–560

Arief VN, DeLacy IH, Crossa J, Payne T, Singh R, Braun HJ, Tian T, Basford KE, Dieters MJ (2015) Evaluating testing strategies for plant breeding field trials: redesigning a CIMMYT international wheat nursery. Crop Sci 55(1):164–177

Arora R (2018) Mechanism of freeze-thaw injury and recovery: A cool retrospective and warming up to new ideas. Plant Sci 270:301–313

Assefa Y, Prasad PVV, Carter P, Hinds M, Bhalla G, Schon R, Jeschke M, Paszkiewicz S, Ciampitti IA (2017) A new insight into corn yield: trends from 1987 through 2015. Crop Sci 57(5):2799–2811

Atlin GN, Cairns JE, Das B (2017) Rapid breeding and varietal replacement are critical to adaptation of cropping systems in the developing world to climate change. Glob Food Sec 12:31–37

Bai YL, Sunarti S, Kissoudis C, Visser RGF, van der Linden CG (2018) The role of tomato WRKY genes in plant responses to combined abiotic and biotic stresses. Front Plant Sci 9:801

Banerjee A, Roychoudhury A (2018) Strigolactones: multi-level regulation of biosynthesis and diverse responses in plant abiotic stresses. Acta Physiol Plant 40(5):86

Bennett D, Reynolds M, Mullan D, Izanloo A, Kuchel H, Langridge P, Schnurbusch T (2012) Detection of two major grain yield QTL in bread wheat (Triticum aestivum L.) under heat, drought and high yield potential environments. Theor Appl Genet 125(7):1473–1485
Bernardo R, Yu J (2007) Prospects for Genomewide Selection for Quantitative Traits in Maize. Crop Science 47(3):1082
Bernardo R (2008) Molecular markers and selection for complex traits in plants: learning from the last 20 years. Crop Sci 48(5):1649–1664
Bernardo R (2010) Breeding for quantitative traits in plants. Stemma Press, Woodbury
Bernardo R (2016a) Bandwagons I, too, have known. Theor Appl Genet 129(12):2323–2332
Bernardo R (2016b) Genomewide predictions for backcrossing a quantitative trait from an exotic to an adapted line. Crop Sci 56(3):1067–1075
Bita CE, Gerats T (2013) Plant tolerance to high temperature in a changing environment: scientific fundamentals and production of heat stress-tolerant crops. Front Plant Sci 4:273
Bloomfield JA, Rose TJ, King GJ (2014) Sustainable harvest: managing plasticity for resilient crops. Plant Biotechnol J 12(5):517–533
Bradshaw JE (2017) Plant breeding: past, present and future. Euphytica 213(3):60
Braun HJ, Rajaram S, vanGinkel M (1996) CIMMYT's approach to breeding for wide adaptation. Euphytica 92(1–2):175–183
Brenner EA, Beavis WD, Andersen JR, Lübberstedt T (2013) Prospects and limitations for development and application of functional markers in plants. In Diagnostics in plant breeding, edited by Lübberstedt, T and R K Varshney, 329–346. Dordrecht: Springer
Campbell BM, Thornton P, Zougmore R, van Asten P, Lipper L (2014) Sustainable intensification: what is its role in climate smart agriculture? Curr Opin Environ Sustain 8:39–43
Canas RA, Yesbergenova-Cuny Z, Simons M, Chardon F, Armengaud P, Quillere I, Cukier C, Gibon Y, Limami AM, Nicolas S, Brule L, Lea PJ, Maranas CD, Hirel B (2017) Exploiting the genetic diversity of maize using a combined metabolomics, enzyme activity profiling, and metabolic modeling approach to link leaf physiology to kernel yield. Plant Cell 29(5):919–943
Castonguay Y, Michaud R, Nadeau P, Bertrand A (2009) An indoor screening method for improvement of freezing tolerance in alfalfa. Crop Sci 49(3):809–818
Cattivelli L (2011) More cold tolerant plants for a warmer world. Plant Sci 180(1):1–2
Chinnusamy V, Schumaker K, Zhu JK (2004) Molecular genetic perspectives on cross-talk and specificity in abiotic stress signalling in plants. J Exp Bot 55(395):225–236
Cooper M, Messina CD, Podlich D, Totir LR, Baumgarten A, Hausmann NJ, Wright D, Graham G (2014) Predicting the future of plant breeding: complementing empirical evaluation with genetic prediction. Crop Pasture Sci 65(4):311–336
Cramer GR, Urano K, Delrot S, Pezzotti M, Shinozaki K (2011) Effects of abiotic stress on plants: a systems biology perspective. BMC Plant Biol 11:163
Crespo-Herrera LA, Crossa J, Huerta-Espino J, Autrique E, Mondal S, Velu G, Vargas M, Braun HJ, Singh RP (2017) Genetic yield gains in CIMMYT's international elite spring wheat yield trials by modeling the genotype x environment interaction. Crop Sci 57(2):789–801
Crossa J, Pérez-Rodríguez P, Cuevas J, Montesinos-López O, Jarquín D,Gustavo de los Campos, Burgueño J, Juan M. González-Camacho, Pérez-Elizalde S,Yoseph Beyene, Susanne Dreisigacker, Ravi Singh, Xuecai Zhang, Manje Gowda, Manish Roorkiwal, Jessica Rutkoski, Rajeev K. Varshney (2017) Genomic Selection in Plant Breeding: Methods, Models, and Perspectives. Trends in Plant Sci 22 (11):961–975
De Santis A, Landi P, Genchi G (1999) Changes of mitochondrial properties in maize seedlings associated with selection for germination at low temperature. Fatty acid composition, cytochrome c oxidase, and adenine nucleotide translocase activities. Plant Physiol 119(2):743–754
Dhillon T, Pearce SP, Stockinger EJ, Distelfeld A, Li CX, Knox AK, Vashegyi I, Vagujfalvi A, Galiba G, Dubcovsky J (2010) Regulation of freezing tolerance and flowering in temperate cereals: the VRN-1 connection. Plant Physiol 153(4):1846–1858
Dolferus R (2014) To grow or not to grow: a stressful decision for plants. Plant Sci 229:247–261

Dubois M, Van den Broeck L, Inze D (2018) The pivotal role of ethylene in plant growth. Trends Plant Sci 23(4):311–323
Dwivedi S, Sahrawat K, Upadhyaya H, Ortiz R (2013) Food, nutrition and agrobiodiversity under global climate change. In: Sparks DL (ed) Advances in agronomy, vol 120. Advances in Agronomy, San Diego, pp 1–128
Dwivedi SL, Ceccarelli S, Blair MW, Upadhyaya HD, Are AK, Ortiz R (2016) Lancrace germplasm for improving yield and abiotic stress adaptation. Trends Plant Sci 21(1):31–42
Dwivedi SL, Scheben A, Edwards D, Spillane C, Ortiz R (2017) Assessing and exploiting functional diversity in germplasm pools to enhance abiotic stress adaptation and yield in cereals and food legumes. Front Plant Sci 8:1461
Eccel E, Rea R, Caffarra A, Crisci A (2009) Risk of spring frost to apple production under future climate scenarios: the role of phenological acclimation. Int J Biometeorol 53(3):273–286
Falloon P, Betts R (2010) Climate impacts on European agriculture and water management in the context of adaptation and mitigation-the importance of an integrated approach. Sci Total Environ 408(23):5667–5687
Feldman M, Levy AA (2009) Genome evolution in allopolyploid wheat-a revolutionary reprogramming followed by gradual changes. J Genet Genomics 36(9):511–518
Fischer RA, Edmeades GO (2010) Breeding and cereal yield progress. Crop Sci 50(2):S85–S98
Foolad MR, Lin GY (2001) Genetic analysis of cold tolerance during vegetative growth in tomato, Lycopersicon esculentum Mill. Euphytica 122(1):105–111
Fracheboud Y, Jompuk C, Ribaut JM, Stamp P, Leipner J (2004) Genetic analysis of cold-tolerance of photosynthesis in maize. Plant Mol Biol 56(2):241–253
Francia E, Barabaschi D, Tondelli A, Laido G, Rizza F, Stanca AM, Busconi M, Fogher C, Stockinger EJ, Pecchioni N (2007) Fine mapping of a HvCBF gene cluster at the frost resistance locus Fr-H2 in barley. Theor Appl Genet 115(8):1083–1091
Francia E, Rizza F, Cattivelli L, Stanca AM, Galiba G, Toth B, Hayes PM, Skinner JS, Pecchioni N (2004) Two loci on chromosome 5H determine low-temperature tolerance in a 'Nure' (winter) x 'Tremois' (spring) barley map. Theor Appl Genet 108(4):670–680
Frascaroli E, Landi P (2013) Divergent selection in a maize population for germination at low temperature in controlled environment: study of the direct response, of the trait inheritance and of correlated responses in the field. Theor Appl Genet 126(3):733–746
Frascaroli E, Landi P (2016) Cold tolerance in field conditions, its inheritance, agronomic performance and genetic structure of maize lines divergently selected for germination at low temperature. Euphytica 209(3):771–788
Frascaroli E, Landi P (2018) Signatures of divergent selection for cold tolerance in maize. Euphytica 214(5):80
Frei OM (2000) Changes in yield physiology of corn as a result of breeding in northern Europe. Maydica 45(3):173–183
Garnett T, Appleby MC, Balmford A, Bateman IJ, Benton TG, Bloomer P, Burlingame B, Dawkins M, Dolan L, Fraser D, Herrero M, Hoffmann I, Smith P, Thornton PK, Toulmin C, Vermeulen SJ, Godfray HCJ (2013) Sustainable intensification in agriculture: premises and policies. Science 341(6141):33–34
Gorjanc G, Gaynor RC, Hickey JM (2018) Optimal cross selection for long-term genetic gain in two-part programs with rapid recurrent genomic selection. Theor Appl Genet 131:1953
Guo Z, Chen D, Schnurbusch T (2018) Plant and floret growth at distinct developmental stages during the stem elongation phase in wheat. Front Plant Sci 9:330
Haak DC, Fukao T, Grene R, Hua Z, Ivanov R, Perrella G, Li S (2017) Multilevel regulation of abiotic stress responses in plants. Front Plant Sci 8:1564
Heffner EL, Lorenz , Jean-Luc Jannink, Mark E. Sorrells, (2010) Plant Breeding with Genomic Selection: Gain per Unit Time and Cost. Crop Science 50 (5):1681
Hensleigh PF, Blake TK, Welty LE (1992) Natural-selection on winter barley composite cross-XXVI affects winter survival and associated traits. Crop Sci 32(1):57–62

Hossain MA, Li ZG, Hoque TS, Burritt DJ, Fujita M, Munne-Bosch S (2018) Heat or cold priming-induced cross-tolerance to abiotic stresses in plants: key regulators and possible mechanisms. Protoplasma 255(1):399–412

Hund A, Frascaroli E, Leipner J, Jompuk C, Stamp P, Fracheboud Y (2005) Cold tolerance of the photosynthetic apparatus: pleiotropic relationship between photosynthetic performance and specific leaf area of maize seedlings. Mol Breed 16(4):321–331

Huner NPA, Dahal K, Kurepin LV, Savitch L, Singh J, Ivanov AG, Kane K, Sarhan F (2014) Potential for increased photosynthetic performance and crop productivity in response to climate change: role of CBFs and gibberellic acid. Front Chem 2:18

Hussain HA, Hussain S, Khaliq A, Ashraf U, Anjum SA, Men SN, Wang LC (2018) Chilling and drought stresses in crop plants: implications, cross talk, and potential management opportunities. Front Plant Sci 9:393

Iraba A, Castonguay Y, Bertrand A, Floyd DJ, Cloutier J, Belzile F (2013) Characterization of populations of turf-type perennial ryegrass recurrently selected for superior freezing tolerance. Crop Sci 53(5):2225–2238

Jermstad KD, Bassoni DL, Wheeler NC, Anekonda TS, Aitken SN, Adams WT, Neale DB (2001) Mapping of quantitative trait loci controlling adaptive traits in coastal Douglas-fir. II. Spring and fall cold-hardiness. Theor Appl Genet 102(8):1152–1158

Jha UC, Bohra A, Jha R (2017) Breeding approaches and genomics technologies to increase crop yield under low-temperature stress. Plant Cell Rep 36(1):1

Josephs EB (2018) Determining the evolutionary forces shaping G*E. New Phytol 219(1):31–36

Junker A, Muraya MM, Weigelt-Fischer K, Arana-Ceballos F, Klukas C, Melchinger AE, Meyer RC, Riewe D, Altmann T (2015) Optimizing experimental procedures for quantitative evaluation of crop plant performance in high throughput phenotyping systems. Front Plant Sci 5:770

Korner C (2016) Plant adaptation to cold climates. F1000Research 5:F1000

Korner C, Basler D, Hoch G, Kollas C, Lenz A, Randin CF, Vitasse Y, Zimmermann NE (2016) Where, why and how? Explaining the low-temperature range limits of temperate tree species. J Ecol 104(4):1076–1088

Kusmec A, Srinivasan S, Nettleton D, Schnable PS (2017) Distinct genetic architectures for phenotype means and plasticities in Zea mays. Nat Plants 3(9):715–723

Lado B, Gonzalez Barrios P, Quincke M, Silva P, Gutierrez L (2016) Modeling genotype x environment interaction for genomic selection with unbalanced data from a wheat breeding program. Crop Sci 56(5):2165–2179

Landi P, Frascaroli E, Lovato A (1992) Divergent full-sib recurrent selection for germination at low-temperature in a maize population. Euphytica 64(1–2):21–29

Li X, Guo T, Mu Q, Li X, Yu J (2018a) Genomic and environmental determinants and their interplay underlying phenotypic plasticity. Proc Natl Acad Sci U S A 115(26):6679–6684

Li X-l, Yang X, Hu Y-x, Yu X-d, Li Q-l (2014) A novel NAC transcription factor from Suaeda liaotungensis K. enhanced transgenic Arabidopsis drought, salt, and cold stress tolerance. Plant Cell Rep 33(5):767–778

Li XH, Wang GH, Fu JJ, Li L, Jia GY, Ren LS, Lubberstedt T, Wang GY, Wang JH, Gu RL (2018b) QTL mapping in three connected populations reveals a set of consensus genomic regions for low temperature germination ability in Zea mays L. Front Plant Sci 9:65

Ly D, Huet S, Gauffreteau A, Rincent R, Touzy G, Mini A, Jannink JL, Cormier F, Paux E, Lafarge S, Le Gouis J, Charmet G (2018) Whole-genome prediction of reaction norms to environmental stress in bread wheat (Triticum aestivum L.) by genomic random regression. Field Crop Res 216:32–41

Makumburage GB, Richbourg HL, LaTorre KD, Capps A, Chen C, Stapleton AE (2013) Genotype to phenotype maps: multiple input abiotic signals combine to produce growth effects via attenuating signaling interactions in maize. G3-Genes Genomes Genet 3(12):2195–2204

Mansouri-Far C, Sanavy S, Saberali SF (2010) Maize yield response to deficit irrigation during low-sensitive growth stages and nitrogen rate under semi-arid climatic conditions. Agric Water Manag 97(1):12–22

Marais DLD, Hernandez KM, Juenger TE (2013) Genotype-by-environment interaction and plasticity: exploring genomic responses of plants to the abiotic environment. Annu Rev Ecol Evol Syst 44(44):5–+

Marla SR, Shiva S, Welti R, Liu SZ, Burke JJ, Morris GP (2017) Comparative transcriptome and lipidome analyses reveal molecular chilling responses in chilling-tolerant sorghums. Plant Genome 10(3):16

Meng LJ, Lin XY, Wang JM, Chen K, Cui YR, Xu JL, Li ZK (2013) Simultaneous improvement in cold tolerance and yield of temperate japonica rice (Oryza sativa L.) by introgression breeding. Plant Breed 132(6):604–612

Meuwissen TH, Hayes EBJ, ME Goddard (2001) Prediction of total genetic value using genome-wide dense marker maps. Genetics 157 (4):1819–1829

Mickelbart MV, Hasegawa PM, Bailey-Serres J (2015) Genetic mechanisms of abiotic stress tolerance that translate to crop yield stability. Nat Rev Genet 16(4):237–251

Montesinos-López OA, Montesinos-López A, Crossa J, Montesinos-López JC, Mota-Sanchez D, Estrada-González F, Gillberg J, Singh R, Mondal S, Juliana P (2018) Prediction of multiple-trait and multiple-environment genomic data using recommender systems. G3: Genes, Genomes, Genetics 8(1):131–147

Mueller D, Schopp P, Melchinger AE (2018) Selection on expected maximum haploid breeding values can increase genetic gain in recurrent genomic selection. G3-Genes Genomes Genet 8(4):1173–1181

Mustafavi SH, Badi HN, Sekara A, Mehrafarin A, Janda T, Ghorbanpour M, Rafiee H (2018) Polyamines and their possible mechanisms involved in plant physiological processes and elicitation of secondary metabolites. Acta Physiol Plant 40(6):102

Naithani S, Preece J, D'Eustachio P, Gupta P, Amarasinghe V, Dharmawardhana PD, Wu G, Fabregat A, Elser JL, Weiser J, Keays M, Fuentes AM-P, Petryszak R, Stein LD, Ware D, Jaiswal P (2017) Plant reactome: a resource for plant pathways and comparative analysis. Nucleic Acids Res 45(D1):D1029–D1039

Nakashima K, Yamaguchi-Shinozaki K, Shinozaki K (2014) The transcriptional regulatory network in the drought response and its crosstalk in abiotic stress responses including drought, cold, and heat. Front Plant Sci 5:170

Nemali KS, Bonin C, Dohleman FG, Stephens M, Reeves WR, Nelson DE, Castiglioni P, Whitsel JE, Sammons B, Silady RA, Anstrom D, Sharp RE, Patharkar OR, Clay D, Coffin M, Nemeth MA, Leibman ME, Luethy M, Lawson M (2015) Physiological responses related to increased grain yield under drought in the first biotechnology-derived drought-tolerant maize. Plant Cell Environ 38(9):1866–1880

Obata T, Witt S, Lisec J, Palacios-Rojas N, Florez-Sarasa I, Yousfi S, Araus JL, Cairns JE, Fernie AR (2015) Metabolite profiles of maize leaves in drought, heat, and combined stress field trials reveal the relationship between metabolism and grain yield. Plant Physiol 169(4):2665–2683

Olesen JE, Borgesen CD, Elsgaard L, Palosuo T, Rotter RP, Skjelvag AO, Peltonen-Sainio P, Borjesson T, Trnka M, Ewert F, Siebert S, Brisson N, Eitzinger J, van Asselt ED, Oberforster M, van der Fels-Klerx HJ (2012) Changes in time of sowing, flowering and maturity of cereals in Europe under climate change. Food Addit Contam Part A Chem Anal Control Expo Risk Assess 29(10):1527–1542

Olesen JE, Trnka M, Kersebaum KC, Skjelvag AO, Seguin B, Peltonen-Sainio P, Rossi F, Kozyra J, Micale F (2011) Impacts and adaptation of European crop production systems to climate change. Eur J Agron 34(2):96–112

Paleari L, Movedi E, Cappelli G, Wilson LT, Confalonieri R (2017) Surfing parameter hyperspaces under climate change scenarios to design future rice ideotypes. Glob Chang Biol 23(11):4651–4662

Presterl T, Ouzunova M, Schmidt W, Moller EM, Rober FK, Knaak C, Ernst K, Westhoff P, Geiger HH (2007) Quantitative trait loci for early plant vigour of maize grown in chilly environments. Theor Appl Genet 114(6):1059–1070

Rapacz M, Ergon A, Hoglind M, Jorgensen M, Jurczyk B, Ostrem L, Rognli OA, Tronsmo AM (2014) Overwintering of herbaceous plants in a changing climate. Still more questions than answers. Plant Sci 225:34–44

Revilla P, Butrón A, Rodríguez VM, Malvar RA, Ordás A (2005) Breeding for cold tolerance. In: Ashraf M, Harris PJC (eds) Abiotic stresses. Plant resistance through breedingand molecular approaches. The Haworth Press, Inc, New York, pp 301–398

Revilla P, Malvar RA, Cartea ME, Ordas A (1998) Identifying open-pollinated populations of field corn as sources of cold tolerance for improving sweet corn. Euphytica 101(2):239–247

Revilla P, Rodriguez VM, Ordas A, Rincent R, Charcosset A, Giauffret C, Melchinger AE, Schon CC, Bauer E, Altmann T, Brunel D, Moreno-Gonzalez J, Campo L, Ouzunova M, Alvarez A, de Galarreta JIR, Laborde J, Malvar RA (2016) Association mapping for cold tolerance in two large maize inbred panels. BMC Plant Biol 16:127

Revilla P, Rodriguez VM, Ordas A, Rincent R, Charcosset A, Giauffret C, Melchinger AE, Schon CC, Bauer E, Altmann T, Brunel D, Moreno-Gonzalez J, Campo L, Ouzunova M, Laborde J, Alvarez A, de Galarreta JIR, Malvar RA (2014) Cold tolerance in two large maize inbred panels adapted to European climates. Crop Sci 54(5):1981–1991

Reynolds M, Langridge P (2016) Physiological breeding. Curr Opin Plant Biol 31:162–171

Rodriguez V, Butron A, Rady MOA, Soengas P, Revilla P (2014) Identification of quantitative trait loci involved in the response to cold stress in maize (Zea mays L.). Mol Breed 33(2):363–371

Salvi S, Tuberosa R (2015) The crop QTLome comes of age. Curr Opin Biotechnol 32:179–185

Sanghera GS, Wani SH, Hussain W, Singh NB (2011) Engineering cold stress tolerance in crop plants. Curr Genomics 12(1):30–43

Seki M, Kamei A, Yamaguchi-Shinozaki K, Shinozaki K (2003) Molecular responses to drought, salinity and frost: common and different paths for plant protection. Curr Opin Biotechnol 14(2):194–199

Semenov MA (2009) Impacts of climate change on wheat in England and Wales. J R Soc Interface 6(33):343–350

Semenov MA, Stratonovitch P, Alghabari F, Gooding MJ (2014) Adapting wheat in Europe for climate change. J Cereal Sci 59(3):245–256

Sezegen B, Carena MJ (2009) Divergent recurrent selection for cold tolerance in two improved maize populations. Euphytica 167(2):237–244

Sheng J, Zheng X, Wang J, Zeng X, Zhou F, Jin S, Hu Z, Diao Y (2017) Transcriptomics and proteomics reveal genetic and biological basis of superior biomass crop Miscanthus. Sci Rep 7:13777

Sthapit BR, Witcombe JR (1998) Inheritance of tolerance to chilling stress in rice during germination and plumule greening. Crop Sci 38(3):660–665

Stojakovic M, Mitrovic B, Zoric M, Ivanovic M, Stanisavljevic D, Nastasic A, Dodig D (2015) Grouping pattern of maize test locations and its impact on hybrid zoning. Euphytica 204(2):419–431

Stone JM, Palta JP, Bamberg JB, Weiss LS, Harbage JF (1993) Inheritance of freezing resistance in tuber-bearing solanum species – evidence for independent genetic-control of nonacclimated freezing tolerance and cold-acclimation capacity. Proc Natl Acad Sci U S A 90(16):7869–7873

Strigens A, Freitag NM, Gilbert X, Grieder C, Riedelsheimer C, Schrag TA, Messmer R, Melchinger AE (2013) Association mapping for chilling tolerance in elite flint and dent maize inbred lines evaluated in growth chamber and field experiments. Plant Cell Environ 36(10):1871–1887

Takeda S, Matsuoka M (2008) Genetic approaches to crop improvement: responding to environmental and population changes. Nat Rev Genet 9(6):444–457

Tampieri E, Baraldi E, Carnevali F, Frascaroli E, De Santis A (2011) The activity of plant inner membrane anion channel (PIMAC) can be performed by a chloride channel (CLC) protein in mitochondria from seedlings of maize populations divergently selected for cold tolerance. J Bioenerg Biomembr 43(6):611–621

Tello-Ruiz MK, Naithani S, Stein JC, Gupta P, Campbell M, Olson A, Wei S, Preece J, Geniza MJ, Jiao Y, Lee YK, Wang B, Mulvaney J, Chougule K, Elser J, Al-Bader N, Kumari S, Thomason J, Kumar V, Bolser DM, Naamati G, Tapanari E, Fonseca N, Huerta L, Iqbal H, Keays M, Fuentes AM-P, Tang A, Fabregat A, D'Eustachio P, Weiser J, Stein LD, Petryszak R, Papatheodorou I, Kersey PJ, Lockhart P, Taylor C, Jaiswal P, Ware D (2018) Gramene 2018: unifying comparative genomics and pathway resources for plant research. Nucleic Acids Res 46(D1):D1181–D1189

Thakur AK, Uphoff NT (2017) How the system of rice intensification can contribute to climate-smart agriculture. Agron J 109(4):1163–1182

Thapa B, Arora R, Knapp AD, Brummer EC (2008) Applying freezing test to quantify cold acclimation in Medicago truncatula. J Am Soc Hortic Sci 133(5):684–691

Thoen MPM, Olivas NHD, Kloth KJ, Coolen S, Huang P-P, Aarts MGM, Bac-Molenaar JA, Bakker J, Bouwmeester HJ, Broekgaarden C, Bucher J, Busscher-Lange J, Cheng X, Fradin EF, Jongsma MA, Julkowska MM, Keurentjes JJB, Ligterink W, Pieterse CMJ, Ruyter-Spira C, Smant G, Testerink C, Usadel B, van Loon JJA, van Pelt JA, van Schaik CC, van Wees SCM, Visser RGF, Voorrips R, Vosman B, Vreugdenhil D, Warmerdam S, Wiegers GL, van Heerwaarden J, Kruijer W, van Eeuwijk FA, Dicke M (2017) Genetic architecture of plant stress resistance: multi-trait genome-wide association mapping. New Phytol 213(3):1346–1362

Tollenaar M, Wu J (1999) Yield improvement in temperate maize is attributable to greater stress tolerance. Crop Sci 39(6):1597–1604

Trachsel S, Burgueno J, Suarez EA, San Vicente FM, Rodriguez CS, Dhliwayo T (2017) Interrelations among early vigor, flowering time, physiological maturity, and grain yield in tropical maize (Zea mays L.) under multiple abiotic stresses. Crop Sci 57(1):229–242

Trnka M, Olesen JE, Kersebaum KC, Skjelvag AO, Eitzinger J, Seguin B, Peltonen-Sainio P, Rotter R, Iglesias A, Orlandini S, Dubrovsky M, Hlavinka P, Balek J, Eckersten H, Cloppet E, Calanca P, Gobin A, Vucetic V, Nejedlik P, Kumar S, Lalic B, Mestre A, Rossi F, Kozyra J, Alexandrov V, Semeradova D, Zalud Z (2011) Agroclimatic conditions in Europe under climate change. Glob Chang Biol 17(7):2298–2318

Tschiersch H, Junker A, Meyer RC, Altmann T (2017) Establishment of integrated protocols for automated high throughput kinetic chlorophyll fluorescence analyses. Plant Methods 13:54

Via S, Gomulkiewicz R, Dejong G, Scheiner SM, Schlichting CD, Vantienderen PH (1995) Adaptive phenotypic plasticity – consensus and controversy. Trends Ecol Evol 10(5):212–217

Viesselmann LM, De Vries B, Dodson HG, Tracy WF (2014) Recurrent selection for seedling growth of sweet corn in cool temperatures. Crop Sci 54(3):1033–1040

Waha K, van Bussel LGJ, Muller C, Bondeau A (2012) Climate-driven simulation of global crop sowing dates. Glob Ecol Biogeogr 21(2):247–259

Xin Z, Browse J (2000) Cold comfort farm: the acclimation of plants to freezing temperatures. Plant Cell Environ 23(9):893–902

Xu WH, Li QX, Jones P, Wang XLL, Trewin B, Yang S, Zhu C, Zhai PM, Wang JF, Vincent L, Dai AG, Gao Y, Ding YH (2018) A new integrated and homogenized global monthly land surface air temperature dataset for the period since 1900. Clim Dyn 50(7–8):2513–2536

Xu Y (2016) Envirotyping for deciphering environmental impacts on crop plants. Theor Appl Genet 129(4):653–673

Zhang F, Ma XF, Gao YM, Hao XB, Li ZK (2014) Genome-wide response to selection and genetic basis of cold tolerance in rice (Oryza sativa L.). BMC Genet 15:55

Zhao YS, Gowda M, Wurschum T, Longin CFH, Korzun V, Kollers S, Schachschneider R, Zeng J, Fernando R, Dubcovsky J, Reif JC (2013) Dissecting the genetic architecture of frost tolerance in Central European winter wheat. J Exp Bot 64(14):4453–4460

Chapter 10
Genetically Engineering Cold Stress-Tolerant Crops: Approaches and Challenges

Rohit Joshi, Balwant Singh, and Viswanathan Chinnusamy

10.1 Introduction

Cold stress is one of the major abiotic stresses that cause significant crop losses in cold climates at high latitudes and altitudes (Janmohammadi et al. 2015). Cold stress primarily affects the survival and geographical distribution of plants. Each year, approximately $2 billion losses occur in global agricultural productivity due to chilling stress. About 15 million ha global cropping area and 7 million ha of rice-cropping area in South and Southeast Asia were lost annually to low-temperature stress (Jha et al. 2017). Similarly, 30% of the wheat-growing area in subtropical, Mediterranean, and temperate regions of Australia witnessed post-head-emergence frost (PHEF) which leads to 85% crop loss annually (Zhao et al. 2015a; Crimp et al. 2016). The negative impacts of cold stress include poor germination, stunted seedling growth, and reduced tillering and fertilization (Maleki and Ghorbanpour 2018). In rice, low temperature causes inhibition/disruption of pollen development resulting in spikelet sterility (Sakata et al. 2014). In addition, cold stress downregulates sugar transport genes in pollen grains leading to reduced starch production (Wani et al. 2016). In chickpea, low temperature inhibits sporogenesis and pollen germination and causes abortion of flower and pod (Kumar et al. 2011).

Plants maintain cellular homeostasis under cold stress by bringing-out changes at molecular, biochemical, and physiological processes. Plants show differential response to chilling (0–15 °C) and freezing (<0 °C) stresses (Maleki and Ghorbanpour 2018).

R. Joshi (✉)
Stress Physiology and Molecular Biology Laboratory, School of Life Sciences, Jawaharlal Nehru University, New Delhi, India

B. Singh
ICAR-National Research Centre on Plant Biotechnology, New Delhi, India

V. Chinnusamy
Division of Plant Physiology, ICAR-Indian Agricultural Research Institute, New Delhi, India

S. H. Wani, V. Herath (eds.), *Cold Tolerance in Plants*,
https://doi.org/10.1007/978-3-030-01415-5_10

Temperate plants are generally chilling tolerant and can acquire freezing tolerance upon prior exposure to suboptimal, nonfreezing temperatures, termed as cold acclimation (Shabala 2017). Extensive studies have been done to decipher the molecular mechanisms of cold acclimation and acquired freezing tolerance in *Arabidopsis* and winter cereals. Few plants require exposure to low temperature for flowering, a process called vernalization (Janmohammadi et al. 2015). Cold stress tolerance is a complex process which involves several metabolic pathways operating at various cellular organelles (Sanghera et al. 2011). Plants reprogram their transcriptome, proteome, and metabolome via intense modifications in their gene expression during cold acclimation (Chinnusamy et al. 2010). A transcriptional network responsible for cold acclimation has been identified that shows the transduction of cold signals from the plasma membrane to the nucleus (Baumann 2017).

Crop improvement for cold tolerance through conventional breeding approach showed little success due to the complexity of stress tolerance traits, lack of precise phenotyping methods and selection criteria, and low genetic variance for cold tolerance in a given breeding population. Genetic engineering for cold tolerance in crops showed that significant increase in cold tolerance can be achieved through engineering transcription factors and other genes involved in signaling (Sanghera et al. 2011). Genomics techniques including a plethora of advanced molecular tools have opened up many exciting possibilities for identification of genes and QTLs involved in stress tolerance and development of crops tolerant to various abiotic stresses including cold stress. Advancement in recombinant DNA technology along with the development of efficient genetic engineering protocols can help to develop precise strategies for generation of cold-tolerant cultivars in various crop species (Wani et al. 2016).

10.2 Cold Stress Sensing and Signaling

Thus far, the sensory network for cold stress response in plants is unresolved. Previous studies showed that plasma membrane is the primary site of damage, i.e., expansion-induced lysis and fracture-jump lesions occur by severe dehydration due to freezing injury in plants. To sense the low temperature and to maintain the vital structure and function of cells, the plasma membrane must remain in its fluid mosaic physical state. However, during cold stress, plasma membrane undergoes phase transitions, from a liquid crystalline to a rigid gel phase with enhanced fatty acid unsaturation (Chinnusamy et al. 2010). It was demonstrated in *Brassica* and alfalfa that low temperature causes membrane rigidification, rearrangement of actin cytoskeleton, and induction of Ca^{2+} channels, resulting in enhanced cytosolic Ca^{2+} levels, which act as second messenger of the cold stress signal (Sangwan et al. 2001). In addition, fad2 mutant impaired in the microsomal oleate Δ12-desaturase or fatty acid desaturase in Arabidopsis showed expression of diacylglycerol (DAG) kinase at higher temperature (18 °C) than the wild-type plants (Vaultier et al. 2006), suggesting the important role of plasma membrane in cold stress sensing. Besides the plasma membrane fluidity, a transmembrane protein COLD1 localized in plasma membrane and endoplasmic reticulum was identified as membrane sensor of cold stress in rice. COLD1 interacts with G-protein α subunit (RGA) and accelerates G-protein GTPase activity to activate the Ca^{2+} channel (Ma et al. 2015). Calcium influx further activates phospholipase C (PLC) and D (PLD), which produce IP_3 and phosphatidic acid, respectively. IP^3 further amplify Ca^{2+} signatures by activating IP_3-gated calcium channels. Loss of function mutants of *FIERY1* (*fry1*) show higher levels of IP^3 with higher induction of *COR* genes and upstream CBF transcription factors (Xiong et al. 2002). Further, calcium exchanger 1 (*cax1*) mutants exhibited higher transcript of *CBF/DREB* proteins and *COR* genes (Catala 2003). Thus, under low temperature, cytosolic Ca^{2+} signatures act as upstream signaling molecule to induce *CBF* transcription factors and *COR* genes (Chinnusamy et al. 2010).

In addition to calcium, ROS also act as signaling molecule to regulate COR genes. *Arabidopsis Frostbite1* (*fro1*) mutant, which is defective in Fe-S subunit (NDUFS4 protein) of complex I of the electron transfer chain showed constitutively higher accumulation of ROS. High ROS accumulation results in decreased expression of *COR* gene and hypersensitivity toward freezing stress (Lee et al. 2002). Cold stress perception in plants results in loss of cell membrane fluidity/rigidification and subsequent transient influx of cytosolic Ca^{2+} causing regulation of COR genes (Chinnusamy et al. 2010). The Ca^{2+} signatures are sensed via Ca^{2+} sensors viz., CaM (calmodulin) (Miura and Furumoto 2013), Ca^{2+}-dependent protein kinase (CDPKs), and CaM-binding transcription activators (CAMTA) (Doherty et al. 2009). Besides calcium sensor proteins, SnRK2 and MAPK families of protein kinases play important role in regulation of C-repeat binding factor (CBF)-dependent or CBF-independent pathways of transcriptional regulation involved in cold acclimation and freezing tolerance (Chinnusamy et al. 2007; Miura and Furumoto 2013; Shi et al. 2015; Zhao et al. 2015a).

10.3 Transcriptional Regulators of Cold Acclimation

During early stages of stress response, various cor genes such as RD29A, COR15A, KIN1, and COR6.6 were found to be induced in *Arabidopsis* (Fig. 10.1). Promoter analysis of COR genes led to the identification of conserved cis-elements C-repeat or dehydration-responsive elements in the promoters of COR genes. Whole-genome analysis of *Arabidopsis* showed more than 10% genes to be involved in cold acclimation (Fowler and Thomashow 2002). Comparative transcriptome analysis of cold-tolerant and cold-sensitive rice genotypes lead to the identification of 242 differentially expressed genes that are involved in senescence, cell death, male sterility, and plant hormone responses (Zhao et al. 2008).

Further, activation of MAP kinase cascade induces inducer of CBF expression1 (ICE1) expression, which activates various kinds of CBF/DREB1 transcription factors belong to ethylene-responsive element-binding factor/APETALA2 (*AP2/DREBP*) family (Jha et al. 2017). These CBFs further induce CRT/DRE-

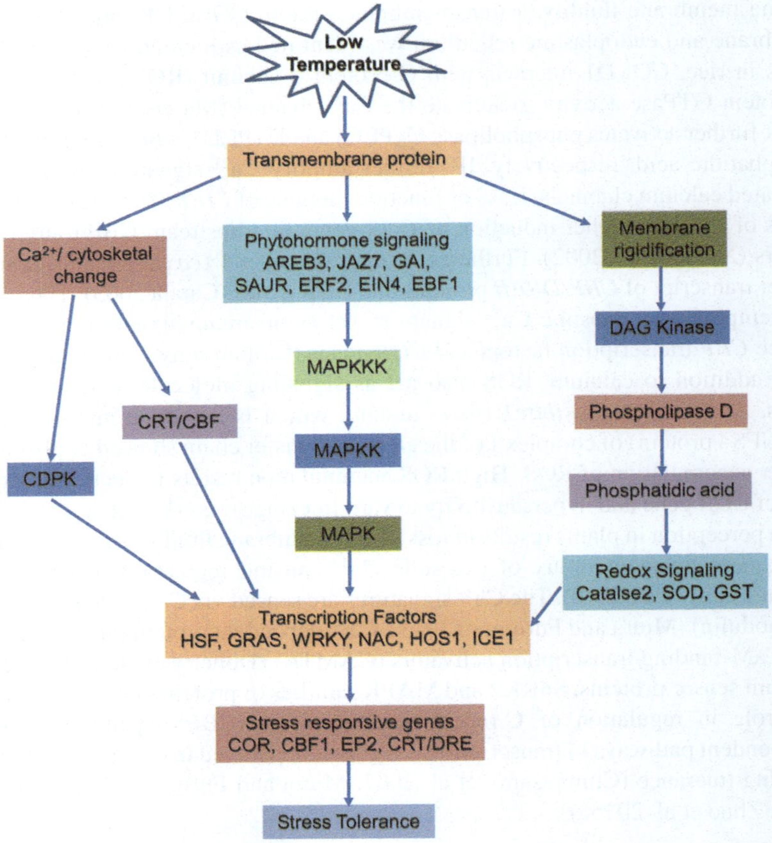

Fig. 10.1 Schematic diagram representing plant signaling under low-temperature stress

regulated downstream target COR genes such as *COR15a* in *Arabidopsis* and *WCS120* in wheat, which encodes cryoprotective proteins (Shi et al. 2015; Zhao et al. 2015a, b; Jha et al. 2017).

Similarly, chaperon proteins like late embryogenesis abundant (LEA) proteins act against cellular damage which indicates their role in anti-aggregation of enzymes under cold stress (Wani et al. 2016). In another study, it has been reported that low-temperature-induced ROS generation results in membrane damage. However, plants showed tolerance mechanism either by increasing antioxidant levels or sugars in apoplastic space or by inducing genes encoding molecular chaperons (Table 10.1). Previous studies have reported synthesis of osmoprotectants such as proline, glycinebetaine, and polyamines and various sugar alcohols such as mannitol, trehalose, and galactinol to protect cell from freezing injury (Sanghera et al. 2011).

Table 10.1 Major cold-responsive genes from different pathways

Gene family	Organism's gene name	Transformed/studied organism	Major function	Reference
Metabolic pathway genes	*Arthrobacter globiformis codA* choline oxidase A	*A. thaliana*	Glycine betaine biosynthesis/ osmoprotectants	Sakamoto et al. (2000)
	prodh proline dehydrogenase	*A. thaliana*	Proline biosynthesis/osmoprotectants	Nanjo et al. (1999)
	CsFFAD fatty acid desaturase	*Cucumis sativus* L.	Fatty acid unsaturation/membrane stabilization	Dong et al. (2016)
	LeFAD7 (omega-3 fatty acid desaturase)	*Lycopersicon esculentum* Mill.	Alleviated the photoinhibition of photo system (PS) 1 and PS2	Liu et al. (2008)
	gpat glycerol-3-phosphate acyltransferase	*N. tabacum*	Fatty acid unsaturation/membrane stabilization	Murata et al. (1992)
	Cu/Zn sod superoxide dismutase of *Manihot esculenta*	*N. benthamiana*	Dismutation of toxic reactive oxygen intermediate/oxidative stress	Xu et al. (2013)
	gst/gpx glutathione S-transferase and glutathione peroxidase	*N. tabacum*	Detoxification of toxic reactive oxygen intermediate/oxidative stress	Roxas et al. (1997)
	EgFAD8	*Elaeis guineensis* Jacq.	Response to light and low temperature	Chen et al. (2018)
	OsTPS1 trehalose-6-phosphate synthase	*Oryza sativa*	Abiotic stress tolerance	Li et al. (2011)
	OsTPP1 trehalose-6-phosphate phoshphatase 1	*Oryza sativa*	Abiotic stress tolerance	Shima et al. (2007)
	Escherichia coli trehalose biosynthetic genes (*otsA* and *otsB*)	*Oryza sativa*	Cold tolerance in rice	Garg et al. (2002)
	Bacterial mannitol-1-phosphate de hydrogenase	*Lycopersicon esculentum* Mill.	Abiotic tolerance	Khare et al. (2010)

(continued)

Table 10.1 (continued)

Gene family	Organism's gene name	Transformed/studied organism	Major function	Reference
Transcription factors	*OsMYBS3*	*Oryza sativa*	Positive regulator of *CBF2/DREB1C*	Su et al. (2010)
	OsbHLH1	*A. thaliana*	Gene regulation and expression	Wang et al. (2003)
	CAMTA3	*A. thaliana*	Positive regulator of *CBF2/DREB1C*	Doherty et al. (2009)
	GmDREB1A (CBF3) DRE-binding protein	*A. thaliana*	Transcription factor	Kidokoro et al. (2015)
	abi3 abscisic acid-induced protein	*A. thaliana*	Transcription factor	Tamminen et al. (2001)
	Osmyb4	*A. thaliana*	Transcription factor	Vannini et al. (2004)
	ZAT12 C2H2 zinc finger	*A. thaliana*	Transcription factor	Vogel et al. (2005)
	mybc1 regulate osmotic stress tolerance	*A. thaliana*	Transcription factor	Zhai et al. (2010)
	ThpI thermal hysteresis proteins (antifreeze protein)	*A. thaliana*	Transcription factor	Zhu et al. (2010)
	OSISAP1 zinc finger protein	*N. tabacum*	Transcription factor	Mukhopadhyay et al. (2004)
	CBF1 CRT/DRE-binding factor 1	*S. lycopersicum*	Transcription factor	Zhang et al. (2011)
Signaling pathway genes	COR genes	Oryza sativa	Negative regulator of drought and cold	Podgorska et al. (2015)
	FIERY1			Shi and Yang (2014)
	RsICE1	*Raphanus sativus*		Man et al. (2017)
	OsbZIP52/RISBZ5	Oryza sativa		Liu et al. (2012)
Transporters	*ala1* aminophospholipid ATPase 1	*A. thaliana*	P-type ATPase (transporter protein)	Chinnusamy et al. (2010)
	Calcium exchanger 1 (cax1)	*A. thaliana*	Defective in a vacuolar Ca2+/H+ antiporter	Chinnusamy et al. (2010)
Posttranscriptionally regulated gene	OsamiRNA319	*O. sativa*		Yang et al. (2013)
	miR159, miR319, miR6022	*S. lycopersicum*		Koc et al. (2015)
	STA1	*A. thaliana*		Lee et al. (2006)
	RCF1	*A. thaliana*		Guan et al. (2013)

10.4 Posttranscriptional Gene Regulation

Posttranscriptional regulation through the processing of pre-mRNA, stabilizing mRNA and export from the nucleus, plays a crucial role during cold acclimation and tolerance (Chinnusamy et al. 2010). Previous studies have identified and functionally characterized various cold-responsive noncoding RNAs (ncRNAs) and their possible targets (Chen et al. 2012; Thiebaut et al. 2012; Niu et al. 2016). Interestingly, these ncRNAs has conservative role in conferring cold stress tolerance in various species. RNA-seq analysis of Yingshuang genotype of tea using Solexa sequencing showed 31 upregulated and 43 downregulated miRNAs, while in Baiye 1 genotype, 46 upregulated and 45 downregulated miRNAs were observed under cold stress (Zhang et al. 2014). Overexpression of OsamiRNA319 that targets OsPCF5 and OsPCF8 genes confer cold stress tolerance in transgenic rice (Yang et al. 2013). Similarly, the role of miR475b in freezing tolerance has been determined in *Populus suaveolens* (Niu et al. 2016). However, in *Solanum lycopersicum* and *S. habrochaites*, various cold-responsive miRNAs were detected such as miR159, miR319, miR6022, miR167, miR169, miR172, miR393, and miR397 (Chen et al. 2015; Koc et al. 2015). Further, in sugarcane, miR319 encoding GAMyb and PCF6 (Thiebaut et al. 2012) and in alfalfa miR156, miR159, miR167, miRNA172, miRNA396, and miRNA398 were found to be cold responsive (Shu et al. 2016). In addition, miR398 targets the Cu/Zn superoxide dismutases (*CSD*) and copper chaperone for SOD (*CCS1*) mRNAs, thus reduction in miR398 expression due to oxidative stress results in enhanced accumulation of CSD1 and CSD2 (Singh et al. 2017).

Arabidopsis COR15A gene which encodes a chloroplast stromal protein having cryoprotective activity plays a key role in providing freezing tolerance in chloroplasts (Nakayama et al. 2007). Likewise, STABILIZED1 (*STA1*), a cold stress-upregulated nuclear pre-mRNA splicing factor, catalyzes the splicing of *COR15A* in *Arabidopsis* and *sta1* mutants and showed hypersensitivity toward cold stress (Lee et al. 2006). Further, alternative splicing of pre-mRNAs of *Arabidopsis* serine/arginine-rich proteins, which are involved in the regulation/execution of mRNA splicing, was observed under cold stress (Palusa et al. 2007). Similarly, *RCF1* (regulator of CBF gene expression1) and *AtRH7/PRH75* genes encoding DEAD-box RNA helicase were found to have significant contribution in assisting pre-mRNA splicing of COR genes and rRNA biogenesis, respectively, in *Arabidopsis* under cold stress (Guan et al. 2013; Huang et al. 2016). Further, Dong et al. (2006) demonstrated the involvement of a nucleoporin AtNUP160 in cold stress tolerance in *Arabidopsis*. Recent reports demonstrated that MAPK signaling modulates cold stress response by phosphorylating ICE1 (Inducer of CBF expression 1) in *Arabidopsis* (Liu and Zhou 2018).

10.5 Epigenetic Regulation During Cold Acclimation

In addition, epigenetic modulations and reprogramming conferring histone modifications includes histone methylation and acetylation, thus conforming its significant role in transcriptional regulation of cold-responsive genes (Jha et al. 2017; Banerjee et al. 2017). However, the expression of epigenetic regulators varies with cold environment and is correlated with the physiological 'memory response' of vernalization (Banerjee and Roychoudhury 2017). Histone deacetylases (HDACs) showed enhanced expression during cold acclimation resulting in deacetylation of the lysine residues on the histone subunits H3 and H4 (Hu et al. 2011). Similarly, in *Medicago truncatula*, the cold stress-induced alternative splicing was observed in the *MtJMJC5* (Jumonji C domain-containing demethylases), which regulates response to the circadian clock (Shen et al. 2016). In addition, HOS1 (High expression of OSmotically responsive gene 1) was found to negatively regulate various cold-responsive genes including CBFs (C-repeat binding factors), CORs, RD (Responsive to Dehydration), KIN, and ICE1 during cold stress (Banerjee et al. 2017). Recently it was demonstrated that low temperature induces WD40-repeat protein, HOS15-mediated chromatin modifications by degrading HD2C which promotes binding of CBFs to the COR gene promoters (Park et al. 2018). HOS15 regulates deacetylation at H4, and *hos15* mutant lines and exhibit cold hypersensitivity with induced transcript of Rd29A (Kim et al. 2015). Further, the activation of heterochromatic tandem-repeat sequence regions, and enhanced acetylation of H3K9ac, plays a significant role during cold acclimation in *Arabidopsis* (Ji et al. 2015), rice (Roy et al. 2014), and maize (Wang et al. 2015).

10.6 Transgenic Approaches to Enhance Cold Stress Tolerance

Inception of genetic engineering techniques in different crops using genes encoding cold-tolerant proteins, metabolites, and traits has opened various exciting arenas for the improvement of cold tolerance. Extensive research efforts revealed a number of transcription factor(s) related to cold stress tolerance, which were isolated and introgressed to delineate the regulatory pathways and to develop improved varieties for enhanced cold tolerance (Wani et al. 2018). However, a single-gene transformation showed a limited success, as cold stress is a multigenic trait. Chilling-tolerant transgenic tobacco (*Nicotiana tabacum*) was developed by the overexpression of omega-3 fatty acid desaturase gene (*FAD7*) under the regulation of cold-inducible promoter that can tolerate 2 °C for 50 days (Khodakovskaya et al. 2006). In another study, chilling tolerance at 1 °C for 11 days was observed in transgenic tobacco plants by transferring a gene encoding a non-specific cyanobacterial desaturase which increase the level of unsaturated fatty acid content in membrane lipids (De Palma et al. 2008). Further, increasing unsaturation of fatty acids by the overexpression of glycerol-3-phosphate acyltransferase (GPAT) also conferred chilling tolerance in rice and tobacco (Wani et al. 2016). Thus, modifying membrane lipid composition can provide cold tolerance by preventing cell membrane leakage.

All the CBFs showed differential expression under cold stress. Three types of *CBF/DREB*1 genes (*CBF3/DREB1a*, *CBF1/DREB1b*, and *CBF2/DREB1c*) are primarily reported to regulate COR gene expression in *Arabidopsis*, and this ICE-CBF-COR signaling pathway is highly conserved among higher eukaryotes (Jha et al. 2017; Wang et al. 2017). In addition, *cbf2* null mutants exhibited enhanced *CBF1* and *CBF3* expression and cold stress tolerance, suggesting that CBF2 is a negative regulator of *CBF1* and *CBF3* (Novillo et al. 2007). Similarly, *TaICE141* and *TaICE187* homologues of ICE1-induced CBF group IV in wheat, and CAMTA1, CAMTA2, and CAMTA3 induced the transcription of CBF1, CBF2, and CBF3, respectively, in *Arabidopsis* under cold stress (Kim et al. 2013). Calmodulin-binding transcription activator 3 (CAMTA3) protein binds to CBF2 promoter and regulates *CBF2* transcription, and *camta3* mutants exhibited low expression of CBF2 as compared to WT plants. Further, *camta1/camta3* double mutants showed hypersensitivity toward cold stress as compared to WT plants (Chinnusamy et al. 2010). ICE1 negatively regulates transcription of MYB15 (an R2R3-MYB family protein) TFs, which negatively regulates downstream CBF genes (Agarwal et al. 2006). Overexpression of *MYB15* in *Arabidopsis* suppresses *CBFs* transcription and cold tolerance, whereas *myb15* T-DNA knockout mutants showed induction of *CBFs* and improved cold tolerance (Agarwal et al. 2006; Xu et al. 2018). Similarly, *ice1* mutants demonstrated enhanced MYB15 expression. Thus, the MYB15-ICE1 interaction appears to play a significant role in modulating CBF transcript during cold acclimation (Chinnusamy et al. 2010). Further, *scream-D* dominant mutant and *ice1* mutant were found to be identical to R236H, and ICE1 protein forms dimer with bHLH transcription factors, *bHLH2*, *bHLH123*, *SPEECHLESS (SPCH)*, *MUTE*, and *FAMA*, which regulate stomatal development (Kanaoka et al. 2008; Yao et al. 2018; Zhao et al. 2018). Similarly, ZAT12 TF also acts as a negative regulator of CBF1, CBF2, and CBF3 during cold stress (Novillo et al. 2007). Overexpression of *ZAT12* in *Arabidopsis* showed reduced *CBFs* transcript under cold stress (Vogel et al. 2005). Likewise, *FIERY2* (*fry2*) mutant showed overexpression of CBFs and COR genes and hypersensitivity toward cold stress, suggesting that *FRY2* (encoding an RNA polymerase II C-terminal domain phosphatase) act as a negative regulator of *CBFs* (Xiong et al. 2002).

In addition, higher expression of osmotically responsive HOS9 (a homeodomain protein) and HOS10 (an R2R3-type MYB) TFs were required for basal cold tolerance (Wani et al. 2016). *hos9* and *hos10 Arabidopsis* mutants showed higher freezing sensitivity than wild type. Likewise, HOS1 mediates ubiquitination and proteolysis of ICE1 and regulates its expression levels in the cell during cold stress (Dong et al. 2006). Overexpression of HOS1 results in substantial reduction in ICE1 protein levels, resulting in hypersensitivity toward freezing stress. HOS10 positively regulates *NCED3* (9-*cis*-epoxycarotenoid dioxygenase) and thus regulates ABA-dependent cold acclimation pathways (Zhu et al. 2005). Among several genes involved in cold stress tolerance and signal transduction, few of them are regulated by C-repeat binding factor/dehydration-responsive element binding (*CBF/DREB*1) transcription factors. Constitutive overexpression of *CBF1/DREB1a* and *CBF3/DREB1b* enhances cold tolerance by inducing cold-regulated (*COR*) genes which leads to the accumulation of sugar and proline and showed tolerance to freezing without any negative effect on growth and development (Jaglo-Ottosen et al. 1998). However, constitutive expression of *CBF1* in tomato and *CBF4* in *Arabidopsis* improved tolerance to cold, salt, and drought stress but showed negative effects like dwarf phenotype and reduction in fruit set and seed number per fruit (Joshi et al. 2016). However, *CBF1* gene overexpressing tomato plants showed higher SOD activity, higher nonphotochemical quenching (NPQ), and lower malondialdehyde (MDA) content, suggesting its role in protecting PSII and PSI during low-temperature stress. Further transcriptome analysis of transgenic tomato and *Arabidopsis* plants overexpressing *LeCBF1* and *AtCBF3* revealed significant differences among the CBF regulons from freezing-tolerant and freezing-sensitive plants (Chinnusamy et al. 2010). These transgenic plants also showed higher induction of the *STZ/ZAT10* zinc finger transcription factor gene, which repressed photosynthesis and carbohydrate metabolism genes, thus reducing growth, which can be reversed by GA_3 treatment (Hsieh et al. 2002a, b). Recently, through genome editing using CRISPR/Cas9 system unravels the role of CBF genes in cold acclimation in *Arabidopsis* (Jia et al. 2016; Zhao et al. 2016), which demonstrate that CBF2 conferred cold tolerance more efficiently than CBF1 and CBF3 (Zhao et al. 2016).

Subsequently, overexpression of *DREB1A* improved cold stress tolerance in tobacco, wheat, and groundnut (Wani et al. 2016). Similarly, expression of a protein phosphatase 2C (*PP2CA*) enhanced cold acclimation and increased freezing tolerance in *Arabidopsis*. In contrast, pyrabactin resistance-like ABA receptors (PYL) inhibit *PP2CA* in both ABA-dependent and ABA-independent manner, and overexpression of *OsPYL3* in *Arabidopsis* improved cold and drought stress tolerance (Lenka et al. 2018). Further, overexpression of Nicotiana Protein Kinase 1 (*NPK1*) provides tolerance to chilling stress (Kovtum et al. 2000). Pennycooke et al. (2003) reported that downregulating α-Gal in petunia results in an increase in freezing tolerance in cold-acclimated plants, whereas overexpression of the α-Gal gene decreases endogenous raffinose and impaired freezing tolerance. The overexpression of *CuCOR19* gene encoding dehydrin (group-2 late embryogenesis abundant (LEA) protein) improved cold tolerance in tobacco (Hara et al. 2003). Likewise, freezing tolerance of *Arabidopsis* was increased by the ectopic expression of the wheat gene *WCS19*, the *Arabidopsis* gene *COR15A*, and coexpression of the genes *RAB18* and *COR47* and *XERO2* and *ERD10* (Sanghera et al. 2011). The freezing tolerance of strawberry leaves was improved after overexpressing wheat dehydrin gene *WCOR410* (Houde et al. 2004). It has also been reported that the expression of cold shock proteins (CspA) from *Escherichia coli* and (CspB) from *Bacillus subtilis* promotes cold stress adaptation in multiple plant species such as rice, *Arabidopsis*, tobacco, and alfalfa (Lopez and Makhatadze 2000). It was demonstrated earlier that

CBFs are positively regulated by ICE1 and negatively regulated by MYB15 (a transcriptional repressor of cold signaling).

Further, mitogen-activated protein kinase (MPK6) phosphorylates MYB15 ($MYB15^{S168A}$), which significantly reduced CBF transcript levels in response to cold stress. Thus, *Arabidopsis* plants overexpressing $MYB15^{S168A}$ are hypersensitive to freezing stress (Kim et al. 2017). In addition, constitutive expression of ring zinc finger protein (*RDCP1*) from hot pepper confers cold stress tolerance in tobacco (Kim et al. 2006). Similarly, overexpression of trehalose-6-phosphate synthase (*TPS*) and trehalose-6-phosphate phosphatase (*TPP*) genes enhanced trehalose accumulation and cold stress tolerance in tobacco and rice (Jang et al. 2003).

Transgenic potato and poplar plants expressing soybean cold-inducible C2H2-type zinc finger transcription factor (SCOF-1) increased cold and freezing stress tolerance in *Arabidopsis* (Kim et al. 2016; Kim 2016). *SCOF-1* interacts with soybean G-box binding factor 1 (*SGBF-1*) and activates ABRE-driven reporter gene expression along with induction of *COR* gene expression (Wani et al. 2016). Mutations in an unknown protein eskimo1 (*ESK1*) increases freezing tolerance (Bouchabke-Coussa et al. 2008). Few other transcription factors such as GmWRKY21 (Zhou et al. 2008), VvWRKY2 (Mzid et al. 2018), TaERF1 (Yi et al. 2004), WLIP19 (Kobayashi et al. 2008), and TCF1 (Ji et al. 2015) also activate the expression of COR genes during cold stress acclimation in soybean, tobacco, *Arabidopsis*, wheat, and *Arabidopsis*, respectively.

10.7 Conclusion and Future Challenges

Various transcriptome studies revealed that cold stress tolerance is a highly complex process. Genetic engineering appears to be an advance and more reliable option than conventional or molecular breeding for the development of improved crop varieties through introgression of numerous selected genes and pyramiding important genes. Significant progress has been made to delineate the molecular mechanism of cold acclimation in various winter crops. Plants reprogram their gene expression employing diverse transcriptional, posttranscriptional, and posttranslational regulatory mechanisms during cold stress. Different overexpression and knockout studies unravel the complex mechanism regulating the quantitative traits such as acquired cold tolerance. This also helps in understanding the complexity of the molecular regulatory networks in crop plants for chilling and freezing tolerance. Transgenic analysis revealed that ICE1-CBF transcriptional cascade plays a crucial role in cold acclimation across different genera. However, few CBF-independent transcriptional pathways were also characterized. Further, epigenetic processes, such as DNA methylation and chromatin modifications, are manifestations of multiple cross talks and plant memory. Moreover, a combined study of physiological, metabolic, and molecular aspects of cold stress tolerance is required to explore the gene regulation leading to whole-plant phenotype under cold stress. In addition, identification of novel cold stress-responsive plasma membrane proteins can also provide an insight into primary stress perception and response.

Acknowledgments Rohit Joshi acknowledges the Dr. DS Kothari Postdoctoral Fellowship from the University Grant Commission, Government of India.

References

Agarwal M, Hao Y, Kapoor A, Dong CH, Fujii H, Zheng X, Zhu JK (2006) A R2R3 type MYB transcription factor is involved in the cold regulation of CBF genes and in acquired freezing tolerance. J Biol Chem 281(49):37636–37645

Banerjee A, Roychoudhury A (2017) Epigenetic regulation during salinity and drought stress in plants: histone modifications and DNA methylation. Plant Gene 11:199–204

Banerjee A, Wani SH, Roychoudhury A (2017) Epigenetic control of plant cold responses. Front Plant Sci 8:1643

Baumann K (2017) Stress responses: membrane-to-nucleus signals modulate plant cold tolerance. Nat Rev Mol Cell Biol 18(5):276–277

Bouchabke-Coussa O, Quashie ML, Seoane-Redondo J, Fortabat MN, Gery C, Yu A, Linderme D, Trouverie J, Granier F, Téoulé E, Durand-Tardif M (2008) ESKIMO1 is a key gene involved in water economy as well as cold acclimation and salt tolerance. BMC Plant Biol 8(1):125

Catala R (2003) Mutations in the Ca^{2+}/H^{+} transporter CAX1 increase CBF/DREB1 expression and the cold-acclimation response in Arabidopsis. Plant Cell Online 15(12):2940–2951

Chen L, Zhang Y, Ren Y, Xu J, Zhang Z, Wang Y (2012) Genome-wide identification of cold-responsive and new microRNAs in Populus tomentosa by high-throughput sequencing. Biochem Biophys Res Commun 417(2):892–896

Chen H, Chen X, Chen D, Li J, Zhang Y, Wang A (2015) A comparison of the low temperature transcriptomes of two tomato genotypes that differ in freezing tolerance: *Solanum lycopersicum* and *Solanum habrochaites*. BMC Plant Biol 15(1):132

Chen L, Wang L, Wang H, Sun R, You L, Zheng Y, Yuan Y, Li D (2018) Identification and characterization of a plastidial ω -3 fatty acid desaturase EgFAD8 from oil palm (Elaeis guineensis Jacq.) and its promoter response to light and low temperature. PLoS One 13(4):e0196693

Chinnusamy V, Zhu J, Zhu J-K (2007) Cold stress regulation of gene expression in plants. Trends Plant Sci 12(10):444–451

Chinnusamy V, Zhu JK, Sunkar R (2010) Gene regulation during cold stress acclimation in plants. In: Plant stress tolerance. Humana Press, New York, pp 39–55

Crimp SJ, Zheng B, Khimashia N, Gobbett DL, Chapman S, Howden M, Nicholls N (2016) Recent changes in southern Australian frost occurrence: implications for wheat production risk. Crop Pasture Sci 67(8):801–811

De Palma M, Grillo S, Massarelli I, Costa A, Balogh G, Vigh L, Leone A (2008) Regulation of desaturase gene expression, changes in membrane lipid composition and freezing tolerance in potato plants. Mol Breed 21(1):15–26

Doherty CJ, Van Buskirk HA, Myers SJ, Thomashow MF (2009) Roles for Arabidopsis CAMTA transcription factors in cold-regulated gene expression and freezing tolerance. Plant Cell Online 21(3):972–984

Dong CH, Hu X, Tang W, Zheng X, Kim YS, Lee BH, Zhu JK (2006) A putative Arabidopsis nucleoporin, AtNUP160, is critical for RNA export and required for plant tolerance to cold stress. Mol Cell Biol 26(24):9533–9543

Dong CJ, Cao N, Zhang ZG, Shang QM (2016) Characterization of the fatty acid desaturase genes in cucumber: structure, phylogeny, and expression patterns. PLoS One 11(3):e0149917

Fowler S, Thomashow MF (2002) Arabidopsis transcriptome profiling indicates that multiple regulatory pathways are activated during cold acclimation in addition to the CBF cold response pathway. Plant Cell 14(8):1675–1690

Garg AK, Kim JK, Owens TG, Ranwala AP, Do Choi Y, Kochian LV, Wu RJ (2002) Trehalose accumulation in rice plants confers high tolerance levels to different abiotic stresses. Proc Natl Acad Sci 99(25):15898–15903

Guan Q, Wu J, Zhang Y, Jiang C, Liu R, Chai C, Zhu J (2013) A DEAD box RNA helicase is critical for pre-mRNA splicing, cold-responsive gene regulation, and cold tolerance in Arabidopsis. Plant Cell 1:tpc-112

Hara M, Terashima S, Fukaya T, Kuboi T (2003) Enhancement of cold tolerance and inhibition of lipid peroxidation by citrus dehydrin in transgenic tobacco. Planta 217(2):290–298

Houde M, Dallaire S, N'Dong D, Sarhan F (2004) Overexpression of the acidic dehydrin WCOR410 improves freezing tolerance in transgenic strawberry leaves. Plant Biotechnol J 2(5):381–387

Hsieh TH, Lee JT, Yang PT, Chiu LH, Charng YY, Wang YC, Chan MT (2002a) Heterology expression of the ArabidopsisC-repeat/dehydration response element binding Factor 1 gene confers elevated tolerance to chilling and oxidative stresses in transgenic tomato. Plant physiology, 129(3):1086–1094

Hsieh TH, Lee JT, Charng YY, Chan MT (2002b) Tomato plants ectopically expressing Arabidopsis CBF1 show enhanced resistance to water deficit stress. Plant Physiology, 130(2):618–626

Hu Y, Zhang L, Zhao L, Li J, He S, Zhou K, Yang F, Huang M, Jiang L, Li L (2011) Trichostatin A selectively suppresses the cold-induced transcription of the ZmDREB1 gene in maize. PLoS One 6(7):e22132

Huang CK, Shen YL, Huang LF, Wu SJ, Yeh CH, Lu CA (2016) The DEAD-box RNA helicase AtRH7/PRH75 participates in pre-rRNA processing, plant development and cold tolerance in Arabidopsis. Plant Cell Physiol 57(1):174–191

Jaglo-Ottosen KR, Gilmour SJ, Zarka DG, Schabenberger O, Thomashow MF (1998) Arabidopsis CBF1overexpression induces COR genes and enhances freezing tolerance. Science 280:104–106

Jang IC, Oh SJ, Seo JS, Choi WB, Song SI, Kim CH, Kim YS, Seo HS, Do Choi Y, Nahm BH, Kim JK (2003) Expression of a bifunctional fusion of the Escherichia coli genes for trehalose-6-phosphate synthase and trehalose-6-phosphate phosphatase in transgenic rice plants increases trehalose accumulation and abiotic stress tolerance without stunting growth. Plant Physiol 131(2):516–524

Janmohammadi M, Zolla L, Rinalducci S (2015) Low temperature tolerance in plants: changes at the protein level. Phytochemistry 117:76–89

Jha UC, Bohra A, Jha R (2017) Breeding approaches and genomics technologies to increase crop yield under low-temperature stress. Plant Cell Rep 36(1):1–35

Ji H, Wang Y, Cloix C, Li K, Jenkins GI, Wang S, Shang Z, Shi Y, Yang S, Li X (2015) The Arabidopsis RCC1 family protein TCF1 regulates freezing tolerance and cold acclimation through modulating lignin biosynthesis. PLoS Genet 11(9):e1005471

Jia Y, Ding Y, Shi Y, Zhang X, Gong Z, Yang S (2016) The cbfs triple mutants reveal the essential functions of CBFs in cold acclimation and allow the definition of CBF regulons in Arabidopsis. New Phytol 212(2):345–353

Joshi R, Wani SH, Singh B, Bohra A, Dar ZA, Lone AA, Pareek A, Singla-Pareek SL (2016) Transcription factors and plants response to drought stress: current understanding and future directions. Front Plant Sci 7:1029

Kanaoka MM, Pillitteri LJ, Fujii H, Yoshida Y, Bogenschutz NL, Takabayashi J, Zhu JK, Torii KU (2008) SCREAM/ICE1 and SCREAM2 specify three cell-state transitional steps leading to Arabidopsis stomatal differentiation. Plant Cell 20(7):1775–1785

Khare N, Goyary D, Singh NK, Shah P, Rathore M, Anandhan S, Sharma D, Arif M, Ahmed Z (2010) Transgenic tomato cv. Pusa Uphar expressing a bacterial mannitol-1-phosphate dehydrogenase gene confers abiotic stress tolerance. Plant Cell Tissue Organ Cult 103(2):267–277

Khodakovskaya M, McAvoy R, Peters J, Wu H, Li Y (2006) Enhanced cold tolerance in transgenic tobacco expressing a chloroplast ω -3 fatty acid desaturase gene under the control of a cold-inducible promoter. Planta 223(5):1090–1100

Kidokoro S, Watanabe K, Ohori T, Moriwaki T, Maruyama K, Mizoi J, Myint Phyu Sin Htwe N, Fujita Y, Sekita S, Shinozaki K, Yamaguchi-Shinozaki K (2015) Soybean DREB1/CBF-type transcription factors function in heat and drought as well as cold stress-responsive gene expression. Plant J 81(3):505–518

Kim YD (2016) Improved cold tolerance by transformation with soybean SCOF-1 gene in *Populus alba*. J Agric Life Sci 50(3):43–53

Kim JS, Park SJ, Kwak KJ, Kim YO, Kim JY, Song J, Jang B, Jung CH, Kang H (2006) Cold shock domain proteins and glycine-rich RNA-binding proteins from Arabidopsis thaliana can promote the cold adaptation process in Escherichia coli. Nucleic Acids Res 35(2):506–516

Kim Y, Park S, Gilmour SJ, Thomashow MF (2013) Roles of CAMTA transcription factors and salicylic acid in configuring the low-temperature transcriptome and freezing tolerance of Arabidopsis. Plant J 75(3):364–376

Kim YS, Lee M, Lee JH, Lee HJ, Park CM (2015) The unified ICE–CBF pathway provides a transcriptional feedback control of freezing tolerance during cold acclimation in Arabidopsis. Plant Mol Biol 89(1–2):187–201

Kim YH, Kim MD, Park SC, Jeong JC, Kwak SS, Lee HS (2016) Transgenic potato plants expressing the cold-inducible transcription factor SCOF-1 display enhanced tolerance to freezing stress. Plant Breed 135(4):513–518

Kim SH, Kim HS, Bahk S, An J, Yoo Y, Kim JY, Chung WS (2017) Phosphorylation of the transcriptional repressor MYB15 by mitogen-activated protein kinase 6 is required for freezing tolerance in Arabidopsis. Nucleic Acids Res 45(11):6613–6627

Kobayashi F, Maeta E, Terashima A, Kawaura K, Ogihara Y, Takumi S (2008) Development of abiotic stress tolerance via bZIP-type transcription factor LIP19 in common wheat. J Exp Bot 59(4):891–905

Koc I, Filiz E, Tombuloglu H (2015) Assessment of miRNA expression profile and differential expression pattern of target genes in cold-tolerant and cold-sensitive tomato cultivars. Biotechnol Biotechnol Equip 29(5):851–860

Kovtum CWL, Tena G, Sheen J (2000) Functional analysis of oxidative stress-activated mitogen activated protein kinase cascade in plants. Proc Natl Acad Sci U S A 97:2940–3005

Kumar S, Malik J, Thakur P, Kaistha S, Sharma KD, Upadhyaya HD, Berger JD, Nayyar H (2011) Growth and metabolic responses of contrasting chickpea (*Cicer arietinum* L.) genotypes to chilling stress at reproductive phase. Acta Physiol Plant 33:779–787

Lee H (2002) LOS2, a genetic locus required for cold-responsive gene transcription encodes a bi-functional enolase. EMBO J 21(11):2692–2702

Lee BH, Kapoor A, Zhu J, Zhu JK (2006) STABILIZED1, a stress-upregulated nuclear protein, is required for pre-mRNA splicing, mRNA turnover, and stress tolerance in Arabidopsis. Plant Cell 18(7):1736–1749

Lenka SK, Muthusamy SK, Chinnusamy V, Bansal KC (2018) Ectopic expression of rice PYL3 enhances cold and drought tolerance in Arabidopsis thaliana. Mol Biotechnol 1:1–2

Li HW, Zang BS, Deng XW, Wang XP (2011) Overexpression of the trehalose-6-phosphate synthase gene OsTPS1 enhances abiotic stress tolerance in rice. Planta 234:1007–1018

Liu Y, Zhou J (2018) MAPping kinase regulation of ICE1 in freezing tolerance. Trends Plant Sci 23(2):91–93

Liu XY, Li B, Yang JH, Sui N, Yang XM, Meng QW (2008) Overexpression of tomato chloroplast omega-3 fatty acid desaturase gene alleviates the photoinhibition of photosystems 2 and 1 under chilling stress. Photosynthetica 46(2):185

Liu C, Wu Y, Wang X (2012) bZIP transcription factor OsbZIP52/RISBZ5: a potential negative regulator of cold and drought stress response in rice. Planta 235(6):1157–1169

Lopez MM, Makhatadze GI (2000) Major cold shock proteins, CspA from Escherichia coli and CspB from Bacillus subtilis, interact differently with single-stranded DNA templates. Biochim Biophys Acta (BBA) Protein Struct Mol Enzymol 1479(1–2):196–202

Ma Y, Dai X, Xu Y, Luo W, Zheng X, Zeng D, Pan Y, Lin X, Liu H, Zhang D, Xiao J, Guo X, Xu S, Niu Y, Jin J, Zhang H, Xu X, Li L, Wang W, Qian Q, Ge S, Chong K (2015) COLD1 confers chilling tolerance in rice. Cell 160(6):1209–1221

Maleki M, Ghorbanpour M (2018) Cold tolerance in plants: molecular machinery deciphered. In: Biochemical, physiological and molecular avenues for combating abiotic stress tolerance in plants. Elsevier Science & Technology, San Diego, pp 57–71

Man L, Xiang D, Wang L, Zhang W, Wang X, Qi G (2017) Stress-responsive gene RsICE1 from Raphanus sativus increases cold tolerance in rice. Protoplasma 254(2):945–956

Miura K, Furumoto T (2013) Cold signaling and cold response in plants. Int J Mol Sci 14(3):5312–5337

Mukhopadhyay A, Vij S, Tyagi AK (2004) Overexpression of a zincfinger protein gene from rice confers tolerance to cold, dehydration, and salt stress in transgenic tobacco. Proc Natl Acad Sci U S A 101:6309–6314

Murata N, Ishizaki-Nishizawa O, Higashi S, Hayashi S, Tasaka Y, Nishida I (1992) Genetically engineered alteration in the chilling sensitivity of plants. Nature 356:710–713

Mzid R, Zorrig W, Ayed RB, Hamed KB, Ayadi M, Damak Y, Lauvergeat V, Hanana M (2018) The grapevine VvWRKY2 gene enhances salt and osmotic stress tolerance in transgenic Nicotiana tabacum. 3 Biotech 8(6):277

Nakayama K, Okawa K, Kakizaki T, Honma T, Itoh H, Inaba T (2007) Arabidopsis Cor15am is a chloroplast stromal protein that has cryoprotective activity and forms oligomers. Plant Physiol 144(1):513–523

Nanjo T, Kobayashi M, Yoshiba Y, Kakubari Y, Yamaguchi-Shinozaki K, Shinozaki K (1999) Antisense suppression of proline degradation improves tolerance to freezing and salinity in Arabidopsis thaliana. FEBS Lett 461(3):205–210

Niu J, Wang J, Hu H, Chen Y, An J, Cai J, Sun R, Sheng Z, Liu X, Lin S (2016) Cross-talk between freezing response and signaling for regulatory transcriptions of MIR475b and its targets by miR475b promoter in *Populus suaveolens*. Sci Rep 6:20648

Novillo F, Medina J, Salinas J (2007) Arabidopsis CBF1 and CBF3 have a different function than CBF2 in cold acclimation and define different gene classes in the CBF regulon. Proc Natl Acad Sci 104(52):21002–21007

Palusa SG, Ali GS, Reddy AS (2007) Alternative splicing of pre-mRNAs of Arabidopsis serine/arginine-rich proteins: regulation by hormones and stresses. Plant J 49(6):1091–1107

Park J, Lim CJ, Shen M, Park HJ, Cha JY, Iniesto E, Rubio V, Mengiste T, Zhu JK, Bressan RA, Lee SY (2018) Epigenetic switch from repressive to permissive chromatin in response to cold stress. Proc Natl Acad Sci U S A 17:201721241

Pennycooke JC, Jones ML, Stushnoff C (2003) Down-regulating α-galactosidase enhances freezing tolerance in transgenic petunia. Plant Physiol 133(2):901–909

Podgorska A, Ostaszewska M, Gardeström P, Rasmusson AG, SZAL B (2015) In comparison with nitrate nutrition, ammonium nutrition increases growth of the frostbite1 Arabidopsis mutant. Plant Cell Environ 38(1):224–237

Roxas VP, Smith RK Jr, Allen ER, Allen RD (1997) Overexpression of glutathione S-transferase/glutathione peroxidase enhances the growth of transgenic tobacco seedlings during stress. Nat Biotechnol 15:988–991

Roy D, Paul A, Roy A, Ghosh R, Ganguly P, Chaudhuri S (2014) Differential acetylation of histone H3 at the regulatory region of OsDREB1b promoter facilitates chromatin remodelling and transcription activation during cold stress. PLoS One 9(6):e100343

Sakamoto A, Valverde R, Chen TH, Murata N (2000) Transformation of Arabidopsis with the codA gene for choline oxidase enhances freezing tolerance of plants. Plant J 22(5):449–453

Sakata T, Oda S, Tsunaga Y, Shomura H, Kawagishi-Kobayashi M, Aya K, Saeki K, Endo T, Nagano K, Kojima M, Sakakibara H (2014) Reduction of gibberellin by low temperature disrupts pollen development in rice. Plant Physiol 1:113

Sanghera GS, Wani SH, Hussain W, Singh NB (2011) Engineering cold stress tolerance in crop plants. Curr Genomics 12(1):30

Sangwan V, Foulds I, Singh J, Dhindsa RS (2001) Cold-activation of Brassica napus BN115 promoter is mediated by structural changes in membranes and cytoskeleton, and requires Ca2+ influx. Plant J 27(1):1–12

Shabala S (ed) (2017) Plant stress physiology. CABI, Boston

Shen Y, Wu X, Liu D, Song S, Liu D, Wang H (2016) Cold-dependent alternative splicing of a Jumonji C domain-containing gene MtJMJC5 in *Medicago truncatula*. Biochem Biophys Res Commun 474(2):271–276

Shi Y, Yang S (2014) ABA regulation of the cold stress response in plants. In: Abscisic acid: metabolism, transport and signaling. Springer, Dordrecht, pp 337–363

Shi H, Qian Y, Tan DX, Reiter RJ, He C (2015) Melatonin induces the transcripts of CBF/DREB1s and their involvement in both abiotic and biotic stresses in Arabidopsis. J Pineal Res 59(3):334–342

Shima S, Matsui H, Tahara S, Imai R (2007) Biochemical characterization of rice trehalose-6-phosphate phosphatases supports distinctive functions of these plant enzymes. FEBS J 274(5):1192–1201

Shu Y, Liu Y, Li W, Song L, Zhang J, Guo C (2016) Genome-wide investigation of MicroRNAs and their targets in response to freezing stress in Medicago sativa L, based on high-throughput sequencing. G3-Genes Genomes Genet 1:g3-115

Singh R, Parihar P, Singh S, Singh MP, Singh VP, Prasad SM (2017) Micro RNAs and nitric oxide cross talk in stress tolerance in plants. Plant Growth Regul 83(2):199–205

Su CF, Wang YC, Hsieh TH, Lu CA, Tseng TH, Yu SM (2010) A novel MYBS3-dependent pathway confers cold tolerance in rice. Plant Physiol 153(1):145–158

Tamminen I, Mäkelä P, Heino P, Palva ET (2001) Ectopic expression of ABI3 gene enhances freezing tolerance in response to abscisic acid and low temperature in Arabidopsis thaliana. Plant J 25:1–8

Thiebaut F, Rojas CA, Almeida KL, Grativol C, Domiciano GC, Lamb CR, Engler JA, Hemerly AS, Ferreira PC (2012) Regulation of miR319 during cold stress in sugarcane. Plant Cell Environ 35(3):502–512

Vannini C, Locatelli F, Bracale M, Magnani E, Marsoni M, Osnato M, Mattana M, Baldoni E, Coraggio I (2004) Overexpression of the rice Osmyb4 gene increases chilling and freezing tolerance of Arabidopsis thaliana plants. Plant J 37:115–127

Vaultier M-N, Cantrel C, Vergnolle C, Justin A-M, Demandre C, Benhassaine-Kesri G, Çiçek D, Zachowski A, Ruelland E (2006) Desaturase mutants reveal that membrane rigidification acts as a cold perception mechanism upstream of the diacylglycerol kinase pathway in cells. FEBS Lett 580(17):4218–4223

Vogel JT, Zarka DG, Van Buskirk HA, Fowler SG, Thomashow MF (2005) Roles of the CBF2 and ZAT12 transcription factors in configuring the low temperature transcriptome of Arabidopsis. Plant J 41(2):195–211

Wang Y-J, Zhang Z-G, He X-J, Zhou H-L, Wen Y-X, Dai J-X, Zhang J-S, Chen S-Y (2003) A rice transcription factor OsbHLH1 is involved in cold stress response. TAG Theor Appl Genet 107(8):1402–1409

Wang Z, Casas-Mollano JA, Xu J, Riethoven JJ, Zhang C, Cerutti H (2015) Osmotic stress induces phosphorylation of histone H3 at threonine 3 in pericentromeric regions of *Arabidopsis thaliana*. Proc Natl Acad Sci 19:201423325

Wang DZ, Jin YN, Ding XH, Wang WJ, Zhai SS, Bai LP, Guo ZF (2017) Gene regulation and signal transduction in the ICE–CBF–COR signaling pathway during cold stress in plants. Biochem Mosc 82(10):1103–1117

Wani SH, Sah SK, Sanghera G, Hussain W, Singh NB (2016) Genetic engineering for cold stress tolerance in crop plants. Adv Genome Sci 4:173–201

Wani SH, Tripathi P, Zaid A, Challa GS, Kumar A, Kumar V, Upadhyay J, Joshi R, Bhatt M (2018) Transcriptional regulation of osmotic stress tolerance in wheat (*Triticum aestivum* L.). Plant Mol Biol 97:469

Xiong L, Schumaker KS, Zhu JK (2002) Cell signaling during cold, drought, and salt stress. Plant Cell 14(suppl 1):S165–S183

Xu J, Duan X, Yang J, Beeching JR, Zhang P (2013) Enhanced reactive oxygen species scavenging by overproduction of superoxide dismutase and catalase delays postharvest physiological deterioration of cassava storage roots. Plant Physiol 161(3):1517–1528

Xu H, Yang G, Zhang J, Wang Y, Zhang T, Wang N, Jiang S, Zhang Z, Chen X (2018) Overexpression of a repressor MdMYB15L negatively regulates anthocyanin and cold tolerance in red-fleshed callus. Biochem Biophys Res Commun 500(2):405–410

Yang C, Li D, Mao D, Liu XU, Ji C, Li X, Zhao X, Cheng Z, Chen C, Zhu L (2013) Overexpression of micro RNA 319 impacts leaf morphogenesis and leads to enhanced cold tolerance in rice (*Oryza sativa* L.). Plant Cell Environ 36(12):2207–2218

Yao P, Sun Z, Li C, Zhao X, Li M, Deng R, Huang Y, Zhao H, Chen H, Wu Q (2018) Overexpression of Fagopyrum tataricum FtbHLH2 enhances tolerance to cold stress in transgenic Arabidopsis. Plant Physiol Biochem 125:85–94

Yi SY, Kim JH, Joung YH, Lee S, Kim WT, Yu SH, Choi D (2004) The pepper transcription factor CaPF1 confers pathogen and freezing tolerance in Arabidopsis. Plant Physiol 136(1):2862–2874

Zhai H, Bai X, Zhu Y, Li Y, Cai H, Ji W, Ji Z, Liu X, Liu X, Li J (2010) A single-repeat R3-MYB transcription factor MYBC1 negatively regulates freezing tolerance in Arabidopsis. Biochem Biophys Res Commun 394:1018–1023

Zhang YJ, Yang JS, Guo SJ, Meng JJ, Zhang YL, Wan SB, He QW, Li XG (2011) Over-expression of the Arabidopsis CBF1 gene improves resistance of tomato leaves to low temperature under low irradiance. Plant Biol 13(2):362–367

Zhang Y, Zhu X, Chen X, Song C, Zou Z, Wang Y, Wang M, Fang W, Li X (2014) Identification and characterization of cold-responsive microRNAs in tea plant (Camellia sinensis) and their targets using high-throughput sequencing and degradome analysis. BMC Plant Biol 14(1):271

Zhao X, Dolferuf R, Darvey N (2008) Precision breeding of cold tolerant rice. IREC Farm Newsl 177:12–13

Zhao C, Lang Z, Zhu JK (2015a) Cold responsive gene transcription becomes more complex. Trends Plant Sci 20(8):466–468

Zhao J, Zhang S, Yang T, Zeng Z, Huang Z, Liu Q, Wang X, Leach J, Leung H, Liu B (2015b) Global transcriptional profiling of a cold-tolerant rice variety under moderate cold stress reveals different cold stress response mechanisms. Physiol Plant 154(3):381–394

Zhao C, Zhang Z, Xie S, Si T, Li Y, Zhu JK (2016) Mutational evidence for the critical role of CBF genes in cold acclimation in Arabidopsis. Plant Physiol 1:00533

Zhao Q, Xiang X, Liu D, Yang A, Wang Y (2018) Tobacco transcription factor NtbHLH123 confers tolerance to cold stress by regulating the NtCBF pathway and reactive oxygen species homeostasis. Front Plant Sci 9:381

Zhou QY, Tian AG, Zou HF, Xie ZM, Lei G, Huang J, Wang CM, Wang HW, Zhang JS, Chen SY (2008) Soybean WRKY-type transcription factor genes, GmWRKY13, GmWRKY21, and GmWRKY54, confer differential tolerance to abiotic stresses in transgenic Arabidopsis plants. Plant Biotechnol J 6(5):486–503

Zhu J, Verslues PE, Zheng X, Lee BH, Zhan X, Manabe Y, Sokolchik I, Zhu Y, Dong CH, Zhu JK, Hasegawa PM (2005) HOS10 encodes an R2R3- type MYB transcription factor essential for cold acclimation in plants. Proc Natl Acad Sci 102(28):9966–9971

Zhu B, Xiong AS, Peng RH, Xu J, Jin XF, Memg XR, Quan-Hong Y (2010) Over-expression of ThpI from Choristoneura fumiferana enhances tolerance to cold in Arabidopsis. Mol Biol Rep 37:961–966

Index

S. H. Wani, V. Herath (eds.), *Cold Tolerance in Plants*,
https://doi.org/10.1007/978-3-030-01415-5

R

S

T

V

The manufacturer's authorised representative in the EU is Springer Nature Customer Service Centre GmbH, Europaplatz 3, 69115 Heidelberg, Germany. If you have any concerns regarding our products, please contact ProductSafety@springernature.com

Printed and bound by CPI Group (UK) Ltd, Croydon, CR0 4YY
30/09/2026
02225229-0003